T0269125

Nutrient management in agricultural watersheds: A wetlands solution

Nutrient management in agricultural watersheds

A wetlands solution

edited by:
E.J. Dunne
K.R. Reddy
O.T. Carton

Wageningen Academic
P u b l i s h e r s

Subject headings:
Wetlands
Water quality
Agriculture

ISBN 9076998612

First published, 2005

Wageningen Academic Publishers
The Netherlands, 2005

Table of contents

4. Wetland biogeochemistry

5. Wetlands, water quality and agriculture

6. Integrated constructed wetlands in Ireland

7. Constructed wetlands and water quality

8. Policy and management of constructed wetland to treat single household and farm-scale wastewaters in Ireland

9. Final discussion

Author index

Keyword index

Organising Committee

K.R. Reddy
Soil and Water Science Department, University of Florida/IFAS, 106 Newell Hall, P.O. Box 110510, Gainesville, FL 32611, USA
Phone: 1-(352)-392-1803
Fax: 1-(352)-392-3399
Email: KRR@ifas.ufl.edu

O.T. Carton
Teagasc Research Centre, Johnstown Castle, Co. Wexford, Ireland
Phone: 353-53-71200
Fax: 353-53-42213
Email: ocarton@johnstown.teagasc.ie

R. Harrington
National Parks and Wildlife Service, Department of Environment, Heritage and Local Government, The Quay, Co. Waterford, Ireland
Phone: 353-51-854329
Email: rharrington@duchas.ie

E.J. Dunne
Wetland Biogeochemistry Laboratory, University of Florida/IFAS, 106 Newell Hall, P.O. Box 110510, Gainesville, FL 32611, USA
Phone: 1-(352)-392-1804
Fax: 1-(352)-392-3399
Email: EJDunne@ifas.ufl.edu

N. Culleton
Teagasc Research Centre, Johnstown Castle, Co. Wexford, Ireland
Phone: 353-53-71200
Fax: 353-53-42213
Email: nculleton@johnstown.teagasc.ie

Sponsors

- Teagasc Research Centre, Wexford, Ireland
- University of Florida/IFAS, USA
    - Soil and Water Science Department
    - Wetland Biogeochemistry Laboratory
- Department of Agriculture and Food, Ireland
- United States Department of Agriculture, USA
- Department of Environment, Heritage and Local Government, Ireland
- Environmental Protection Agency, Ireland
- Glanbia, Ireland

Introduction

An international symposium on *"Nutrient Management in Agricultural Watersheds: A Wetlands Solution,"* was organised by Teagasc Research Centre, Johnstown Castle, Wexford, Ireland and the Wetland Biogeochemistry Laboratory, Soil and Water Science Department, University of Florida/IFAS, Gainesville, Florida. This internationally collaborated symposium was co-sponsored by the Department of Agriculture and Food, Ireland; Department of Environment, Heritage and Local Government, Ireland; Environmental Protection Agency, Ireland; and the United States Department of Agriculture.

Diversity and subsequent change in agricultural land use is an integral component of European agri-environmental policy. Changes are not only required, but they are ever present; converting agricultural land to wetland for improvement of water quality is only one illustration of this change. Addressing water quality issues within agricultural watersheds requires multidisciplinary approaches, therefore a more holistic approach and the integration of skills from multiple sectors and disciplines is required to counteract the problem of contaminant and nutrient loss from agriculture.

The purpose of this symposium was to provide a forum for the synthesis of current information on the use of wetlands both constructed and natural that are increasingly used to reduce point and non-point source nutrient and contaminant loss from agricultural practices. International /national insights and experiences were presented for basic information transfer to stakeholders and to illustrate the interdisciplinary nature of water resource management in agricultural watersheds and the use of wetlands. Future directions for both research and management were also identified. Finally, it was apparent that polarization of views among stakeholders was present. Such views may be useful for establishing points of view; however these may not be beneficial to achieving general consensus so many stakeholders claim they want.

The symposium covered aspects such as: water quality issues in agricultural watersheds; fundamental functions and values of wetlands within agricultural watersheds; present conventional management practices to reduce nutrient and contaminant loss from agriculture; some policy and regulatory issues relating to water resource management; wetlands and water quality; wetland restoration; overview of nitrogen and phosphorus cycling within wetland ecosystems; and finally, some wetland case studies were presented.

During the symposium a total of 26 oral presentations and 12 poster papers were presented. Contributors to this international symposium were from several countries (Ireland, Northern Ireland, U.K., Czech Republic, Norway, Spain, and USA). Symposium participants came from a range of disciplines and sectors including state bodies, research centres, universities, consultants, county councils, non-governmental organisations, land managers and farmers.

Special thanks are extended to our sponsors and participants (oral presenters, poster presenters, session moderators and registrants). The assistance of the University of Florida/IFAS, Office of Conferences and Institutes staff, especially Ms. Sharon Borneman and Ms. Eleanor Spillane, Teagasc Research Centre, Johnstown Castle, Wexford who handled most of the symposium arrangements and registration, are also gratefully acknowledged. We also thank Dr. Austin O'Sullivan, Mr. Adrian Stafford, Mr. John Sinnot, Mrs. Mary O'Sullivan, Mr. Micheal Quirke, Ms. Sarah Lacey, and Ms. Susan O'Neill who helped out during the symposium. The significant contribution of Ed. Dunne to the organisation of the symposium is also acknowledged.

The Organising Committee
K.R. Reddy
O.T. Carton,
E.J. Dunne
R. Harrington
N. Culleton

Water quality in Ireland - diffuse agricultural eutrophication - a key problem

M. McGarrigle
Environmental Protection Agency, John Moore Road, Castlebar, Co. Mayo, Ireland

The Irish Environmental Protection Agency (EPA) and its predecessor organisations have monitored Irish river water quality since the early 1970s. Over 3,200 separate locations on Irish rivers are routinely monitored, covering some 13,200 km of main-stem river channel. This enables the extent of river pollution to be estimated and trends in water quality with time to be determined (Figure 1). Serious organic pollution has declined to approximately one percent of river channels surveyed at present. However, eutrophication has increased during the same time period and almost 30 percent of river channel now suffers from eutrophication (McGarrigle *et al.*, 2002).

Almost half of all observed river pollution is now due to agricultural sources. The monitoring network density (average distance between monitoring points on main river channel is less than 5 km) allows for a detailed appraisal of pollution sources. Point sources are investigated directly in the field. In the case of diffuse pollution, point sources are first eliminated as the potential cause and changing land use practices are linked to changes in water quality using statistical models. Furthermore, recent studies in Ireland provide strong evidence that higher soil phosphorus (P) levels greatly increase the risk of P loss from soil to water (Daly *et al.*, 2002; Jordan *et al.*, In Press; Scanlon *et al.*, 2004; Scanlon *et al.*, In Press). Reducing diffuse P loss is now regarded as an essential element of successful river pollution control.

Irish lakes and rivers are naturally salmonid with sustainable populations of salmon, trout and arctic char, requiring good water quality. Thus, for the protection of these waters eutrophication control is a high priority. However, of the 304 lakes assessed in the 1998-2000 period, 44 were eutrophic. The eutrophication of Lough Sheelin in County Cavan provides an important example of the decline of a large wild brown trout fish population resulting from pig slurry runoff from grassland areas in a catchment dominated by impermeable gley soils. Furthermore, many smaller Irish lakes have become eutrophic in the relatively recent past. Recent evidence also demonstrates that P can reach groundwater directly accompanied by ammonia and pathogen contamination in vulnerable zone areas, as well as high nitrate concentrations reaching groundwater. Some 38 percent of aquifer samples taken by the EPA were contaminated by faecal coliform bacteria indicating direct gross organic contamination.

Estuarine and coastal waters are also affected by eutrophication due to phosphorus and nitrogen with some 13 of 47 Irish estuaries and bays examined found to be eutrophic.

If eutrophication is to be brought under control it is crucial to balance inputs and outputs of P on farms. Nutrient management planning on all farms is essential in order to reduce the unnecessary P loss to water. Control of farmyard pollution is also critical, but there is a growing body of evidence to suggest that diffuse sources may contribute even more P to water than do farmyards. An integrated strategy requires control of both. The recent State of the

Environment Report for Ireland (EPA 2004) emphasises the importance of controlling diffuse sources of P.

In conclusion, the European Union Water Framework Directive (200/60/EC) legislatively requires Member States to achieve "good ecological status" for all waters by 2015. Thus, it is important to quantify the relative importance of both diffuse and point source pollution from agriculture with ongoing reductions in agricultural P loss to water required. Finally, the integrated management of water resources at the watershed-scale is necessary to achieve good ecological status for all Irish waters by 2015.

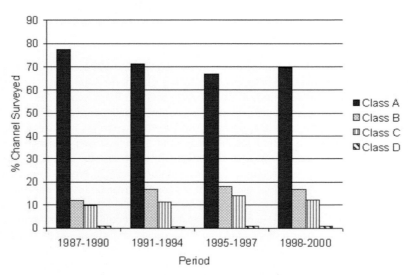

Figure 1. Trends in Irish river water quality 1987-2000. Class A indicates generally satisfactory quality or unpolluted conditions; Class B is slightly polluted; Class C is moderately polluted; and Class D indicates serious pollution.

References

Daly, K., P. Mills, B. Coulter, and M. McGarrigle, 2002. Modelling phosphorus concentrations in Irish rivers using land use, soil type and soil phosphorus data. Journal of Environmental Quality **31** 590-599.

Environmental Protection Agency, 2004. Ireland's Environment 2004. Environmental Protection Agency, Wexford, Ireland.

Jordan, P., W. Menary, K. Daly, G. Kiely, G. Morgan, P. Byrne, and R. Moles. In press. Patterns and processes of phosphorus transfer from Irish grassland soils to rivers - integration of laboratory and catchment studies. Journal of Hydrology.

McGarrigle, M.L., J.J. Bowman, K.C. Clabby, J.L. Lucey, P. Cunningham, M. MacCárthaigh, M. Keegan, B. Cantrell, M. Lehane, C. Clenaghan, and P. Toner, 2002. Water Quality in Ireland 1998-2000. Environmental Protection Agency, Wexford, Ireland.

Scanlon, T.M. G. Kiely, and Q. Xie, 2004. A nested catchment approach for defining the hydrological controls on non-point phosphorus transport. Journal of Hydrology **291** 218-231.

Scanlon, T.M., G. Kiely, and R. Amboldi, In Press. Model determination of non-point source phosphorus transport pathways in a fertilized grassland catchment. Hydrological Processes.

Soil hydrology as a factor in diffuse phosphorus pollution

C. O'Reilly[1,2], W.L. Magette[2], K. Daly[1] and O.T. Carton[1]
[1]Teagasc Environmental Research Centre, Johnstown Castle, County Wexford, Ireland
[2]Biosystems Engineering Department, University College Dublin, Earlsfort Terrace, Dublin 2, Ireland

Abstract

The hydrological pathways responsible for transporting phosphorus (P) from soil to waterways are overland flow, interflow (shallow subsurface flow between the soil surface and A/B horizon) and groundwater flow. The extent to which these pathways occur is primarily dependent upon precipitation, soil physical and chemical characteristics, soil moisture status and land management. An accurate field method, using a rainfall simulator designed to mimic the characteristics of natural rainfall in south east Ireland, and allowing for detailed analysis of overland and subsurface flow is being carried out to provide an accurate method to measure the hydrologic responses of different soils and the transport of P there from.

Keywords: phosphorus, rainfall simulator, interflow, overland flow.

Introduction

The EPA in Ireland has shown that nutrient loss from agriculture is now the biggest source of phosphorus (P) pollution in Irish rivers (EPA, 2001). Increasingly intensive farming practices are believed to be responsible for the excess nutrient losses that are the main cause of eutrophication in Ireland's waterways. Tunney *et al.* (2000) showed that the increase in fertiliser use over the past 50 years has resulted in a tenfold increase in soil test P levels. Effective environmental management of P losses requires information on how the source and transport vector (in this study, water) interact. Preventing P loss could then concentrate on defining and managing source-areas of P (those physical areas having high soil test phosphorus (STP) concentrations and/or receiving large inputs of P) that coincide with high overland flow and/or subsurface flow.

Two problems typically inhibiting research to define the hydrologic responses of soils are the cost of investigations and the fact that natural rainfall can neither be precisely predicted nor controlled. These obstacles can be partially overcome using a rainfall simulator to create rainfall that has predetermined characteristics. Simulated rainfall allows for control of intensity (in terms of duration), drop size, wind interference and overall data collection. By carefully measuring rainfall together with the quantity and quality of P transport pathways such as overland flow and shallow subsurface flow, it is possible to identify the interactions between soil hydrology and P transport from the landscape.

The aim of this research is to (i) quantify and subsequently explain the hydrological responses, i.e. infiltration, overland flow and subsurface flow, of selected soils to simulated rainfall and, (ii) to quantify the influence of these hydrologic responses on P losses occurring from the soil. This research will also assess the contribution that this method may have towards classifying Irish soils according to their hydrologic characteristics.

Materials and methods

Design specifications were obtained from USDA Columbia Plateau Research Centre, Pendleton, Oregon. Rainfall in the Pacific Northwest region of USA is of low intensity and long duration, similar to Irish conditions. The rainfall simulator consists of three identical rotating disc nozzle assemblies. Each rotating disc-nozzle assembly consists of an inlet pipe, on the end of which is a nozzle, a gearmotor which drives a rotating disc, and a catchpan to return excess water. Each disc has on it four equally sized open slots that can be covered by inserting pans, thereby allowing rainfall to be simulated at four different intensities. A petrol generator supplies power to the gearmotors and water pump.

Rainfall was applied from a height of 2.5 m to a test plot measuring 9 m (down slope length) by 1.5m (width). The plot was hydrologically isolated in relation to runoff collection using stainless steel plot borders, 1.1 m in length and inserted 0.15 m into the soil. The plot was hydrologically isolated in relation to interflow by digging a trench around the outside of the steel plot borders to a depth of 40 cm and backfilling with a compacted gleyed soil. At the lower end of the plot, two separate collector pipes, one at the soil surface, the other level with the A/B horizon, collect surface runoff and interflow respectively. Pipes deliver the flow to a point where intermittent sampling can be carried out and flow rates can be measured using a measuring cylinder and stopwatch.

Water is obtained from the mains supply and filtered through an ion-exchange column into a 1500 litre tank. The tank provides enough water to produce uninterrupted rainfall for between three and four hours depending on the intensity at which the rainfall is applied. A system of hoses, gate valves and pressure gauges was developed to ensure a consistent water pressure supply to each of the three simulator units, from a single-outlet water pump. Soil moisture within the plot is recorded using TDR moisture probes connected to the datalogger.

Land management and use of the sites under investigation has been controlled over the twelve months previous to testing. Some soil physical and chemical properties of each site were measured (Table 2). Overland flow and interflow are sampled at 10 minute intervals after initiation and tested in the laboratory for total P (TP), total dissolved P (TDP), dissolved reactive P (DRP), and total reactive P (TRP) within 24 hours.

Calibration of the rainfall simulator involved drop size determination, pressure transducer calibration and uniformity of distribution coefficient. The flour pellet method of Laws and Parsons (1943) was used to determine drop size. The Christiansen Coefficient of application uniformity (Christiansen, 1942), Cu, was determined at the four different rainfall intensities. It was measured using 125 evenly spaced tins within the plot. Rainfall is simulated for a fixed time period and the water volume in each can is measured and converted to a depth value (mm h^{-1}). Results are tabulated and applied to the following formula, the distribution being perfect when Cu is equal to 100.

$$Cu = \left[\frac{1 - \text{average deviation from mean}}{\text{mean depth of water}}\right] * 100 \qquad (1)$$

Results

The results of the Christiansen coefficient tests are shown in Table 1. The effect of overlap between the rainfall produced by the three rotating disc assemblies became most apparent at the lowest intensity. The highest intensity of 11.2 mm hr^{-1} has a return period of one year according to Johnstown Castle climate data provided by Met Eireann (2004).

An illustration of the rainfall distribution when one pan was inserted into each rotating disc nozzle assembly is shown in Figure 1. The rotating disc nozzle assemblies are positioned 1.5m, 4.5m and 7.5m from the base of the plot. Each data point on the graph represents the mean of 5 measurements taken per row.

Table 2 provides a description of each of the two sites using data from tests already carried out. The sites have a similar history in terms of agronomic use over the previous twelve months and differences in drainage class should give differing hydrological responses.

Table 1. Mean rainfall rates and corresponding Christiansen coefficients.

Number of pans	Mean rainfall rate (mm hr^{-1})	Cu
0	11.2	70
1	8.2	70
2	5.7	70
3	2.4	63

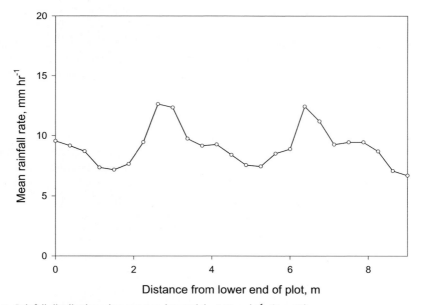

Figure 1. Rainfall distribution when one pan inserted (μ=8.2mm hr^{-1}, Cu = 70).

Table 2. Site physical and chemical characteristics.

Property	Site 1	Site 2
Drainage Class	Moderate/Well	Imperfect
Slope (%)	3.0	3.0
Soil Group	Brown Earth	Gley
Soil Texture of A horizon	Sandy Loam	Sandy Loam
Soil Texture of B horizon	Clay loam	Clay loam
Depth to A/B horizon (cm)	30	35
% Clay	15.1	17.1
% Silt	26.5	30.2
% Sand	58.4	52.7
Bulk Density (g cm^{-3})	0.89	1.04
Particle Density (g cm^{-3})	2.25	2.32
Porosity	0.60	0.55
% Organic matter	9.7	8.4
Morgan's P (mg kg^{-1})	21.5	29.0
Teagasc soil P Index	2	2
pH	6.1	6.0

Discussion

To date, all calibration tests have been completed and suggest that the simulator is extremely accurate and a reliable tool for conducting repeatable hydrologic experiments under field conditions. Relationships will be established for hydrological responses, and the resulting transport of phosphorus, at different initial soil moisture levels and different rainfall intensities within each site. Further soil analysis, e.g. lateral and vertical hydraulic conductivities, should aid explanation of differences in hydrologic response between sites.

Conclusions

The results from this research will provide important information on such factors as the times that waterways are most susceptible to P pollution and how influential soil type, soil moisture status and climatic factors are. The findings should not only provide an enhanced theoretical description of runoff / P transport interactions, but also aid the determination of additional steps in protecting Irish waterways, by providing an improved basis for agricultural P management guidelines.

Acknowledgements

This project is funded by the Teagasc Walsh Fellowship scheme in association with University College Dublin.

References

Christiansen, J.R., 1942. Irrigation by sprinkling. Bulletin 670. University of California - Davis, Davis CA.

Environmental Protection Agency, 2001. Developing a national phosphorus balance for agriculture in Ireland. Published by Environmental Protection Agency, County Wexford, Rep. of Ireland, pp. 1-14.

Laws, J.O. and D.A. Parsons, 1943. The relationship of raindrop-size to intensity. Transactions of the American Geophysical Union **24** 452-460.

Met Eireann, 2004. Personal Communication. Met Eireann, The Irish Meteorological Service, Glasnevin Hill, Dublin 9, Ireland.

Tunney H., B. Coulter, K. Daly, I. Kurz, C. Coxon, D. Jeffrey, P. Mills, G. Kiely, and G. Morgan, 2000. Quantification of phosphorus loss to water due to soil P desorption. Final report and literature review. Environmental Protection Agency, Johnstown Castle, Wexford, Ireland.

Flow effects on phosphorus loss in overland flow

D. Doody[1,2], I. Kurz[1] H. Tunney[1] and R. Moles[2]
[1]Teagasc Research Centre, Johnstown Castle, Wexford, Ireland
[2]Department of Chemical and Environmental Sciences, University of Limerick, Limerick, Ireland

Abstract

Field studies have demonstrated an increase in dissolved reactive phosphorus (DRP) concentration in overland flow with an increase in overland flow rate. This is counter-intuitive, as due to dilution, a decrease in P concentration would be expected. This research investigates the impact of the expansion of a variable source area (VSA) during a rainfall event on DRP concentration in overland flow. Intact soil sods were taken from an agricultural grassland soil and placed within a 2.1 m long laboratory flume. The laboratory flume was used to simulate overland flow and the expansion of a VSA. The results demonstrated that as flow path length increased there was a corresponding increase in DRP concentration in overland ($p < 0.01$). An increase in flow rate resulted in a decrease in DRP concentration in overland flow ($p < 0.01$). When flow path length was increased in conjunction with flow rate DRP concentration increased despite the impact of dilution.

Keywords: phosphorus, variable source area, flowpath, flow-rate, flume.

Introduction

Phosphorus (P) loss from soil to water is controlled by source and transport factors (McDowell *et al.*, 2001). Transport factors provide the energy and the carrier in which potential mobile P (PMP) is transported from the source to lakes and rivers (Haygarth and Jarvis, 1999). Although significant quantities of P have been reported in subsurface flow (Sims *et al.*, 1998), overland flow is the main pathway of P loss from agricultural soil (Sharpley *et al.*, 1994).

Investigations into P loss from soil to water at field-scale have demonstrated that as overland flow rate increases, there is a corresponding increase in the concentration of DRP recorded in overland flow (Figure 1) (Kurz, 2002). This is counter-intuitive, as due to dilution a decrease in DRP concentration would be expected. The factors controlling this relationship have not yet been identified.

In many catchments relatively small well-defined areas of saturation are responsible for the majority of overland flow generated (Pionke *et al.*, 2000) and these areas are referred to as variable source areas (VSAs) (Gburek and Sharpley, 1998). VSAs occur in areas where the water table is close to the soil surface or in areas of high soil moisture. Where VSAs intersect areas of the landscape that have large quantities of P, critical source areas (CSAs) are created that can account for 90% of P exported from a catchment (Gburek and Sharpley, 1998). Vulnerable source areas are dynamic, expanding and contracting rapidly during rainfall events (Gburek and Sharpley, 1998).

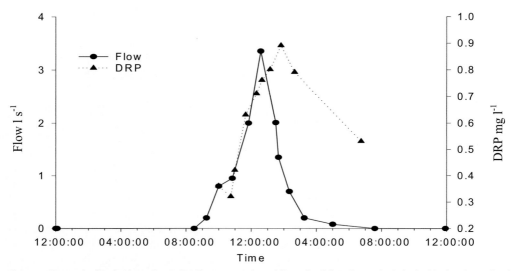

Figure 1. Change in dissolved reactive P (DRP) concentration with overland flow from a hydrological isolated grassland field site (Kurz, 2002).

The objective of this research was to investigate the impact of variations in overland flow path length, caused by the expansion of VSAs, and flow rate on DRP concentration in overland flow. This was carried out under simulated overland flow conditions

Materials and methods

Intact grassland soil sods, of dimensions 2.1 x 0.05 m were taken from a pasture, which had a soil test P (STP) value of 17 mg l^{-1} (Morgan's P). These sods were then placed within a laboratory flume (Figure 2). Overland flow was simulated using a peristaltic pump and a mobile sprinkler system to allow for the movement of the water source up and down the length of the grass sod. To simulate a non-expanding VSA, water was pumped into the top of the flume and the flow path length was held steady throughout the duration of the experiment. An expanding VSA was simulated by the water source being incrementally moved up the length of the flume throughout the duration of the experiment. The impact of three flow rates of 44 ml min^{-1}, 88 ml min^{-1} and 132 ml min^{-1} were also investigated. The experiments were carried out for six hours with water samples collected every 30 minutes, filtered and analysed colormetrically for DRP (Murphy and Riley, 1962). Data was analysed using multiple linear regression.

Results and discussion

When the flow path length was held steady throughout the experiment, DRP concentration decreased rapidly due to dilution (Figure 3). However, when the flow path length was increased incrementally throughout the experiment, DRP concentration increased with flow path length ($p < 0.01$) (Figure 3). The differences in the trends of the DRP concentrations from these two

Figure 2. Laboratory flume containing intact grassland soil sods.

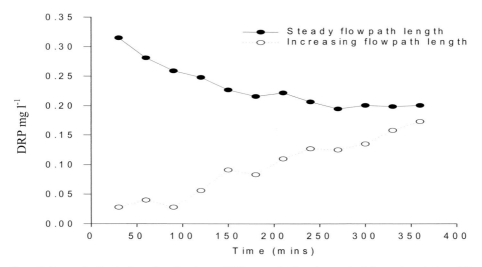

Figure 3. Change in dissolved reactive phosphorus (DRP) concentration at a constant flow rate under two different flow path regimes.

experiments may be due to the availability of freely desorbable P (McDowell and Sharpley, 2003; Vadas and Sims, 2002). During the steady flow path experiment, DRP concentration decreased as the pool of freely desorbable P was depleted. However, it is likely that when the flow path length was increased throughout the experiment, new areas of freely desorbable P were constantly accessed and so, no depletion of source may have occurred. With an increasing flow path length, dilution also had less of an impact on DRP concentration, as the area over which the overland was travelling constantly increased. An increase in flow rate resulted in a decrease in DRP concentration (p <0.01) (Figure 4). When flow rate was increased in conjunction with an increase in flow path length, DRP concentration continued to increase as long as the flow path length was increased (Figure 5). This suggests that flow path length has a more dominant impact on dissolved P concentration than flow rate.

The results from this study suggest that the increase in DRP concentration with an increase in overland flow rate is controlled by VSA hydrology at bench-scale. Further work is required to test this hypothesis at field-scales.

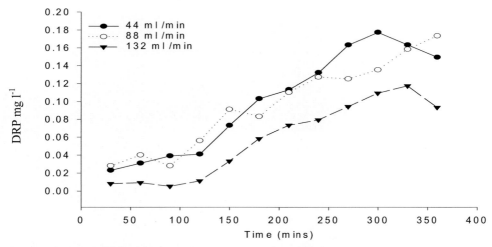

Figure 4. Variation in dissolved reactive phosphorus (DRP) concentration at three different flow rates with increasing flow path length.

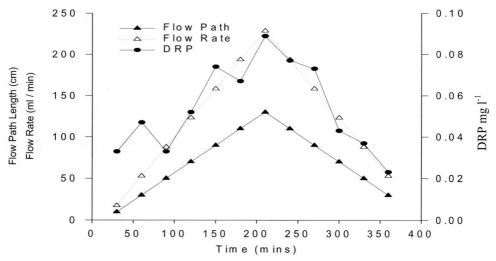

Figure 5. Variation in dissolved reactive phosphorus (DRP) concentration with variable flow path length and variable flow rate.

Conclusions

- An increase in flow path length resulted in an increase in DRP concentration.
- An increase in flow rate caused a decrease in DRP concentration.
- When an increase in flow path length is accompanied by an increase in flow rate, DRP concentration increases despite the impact of dilution.
- VSA hydrology should be investigated at field-scales to examine its role in the increase in DRP concentration with an increase in overland flow rate.

Acknowledgements

This research was funded by the EPA (Ireland) and Teagasc.

References

Gburek, W. and A.N. Sharpley, 1998. Hydrological controls on phosphorus loss from upland agricultural watersheds. Journal of Environmental Quality **27** 267-277.

Haygarth, P. and S. Jarvis, 1999. Transfer of phosphorus from agricultural soils. Advances in Agronomy **66** 195-247.

Kurz, I., 2002. Phosphorus loss from agricultural grassland in overland flow and subsurface drainage water. Unpublished Ph.D. Thesis, Dublin University, Ireland.

McDowell, R. and A.N. Sharpley, 2003 . Phosphorus solubility and release kinetics as a function of soil test P concentration. Geoderma **112** 143-154.

McDowell, R., A.N. Sharpley and G. Folmar, 2001. Phosphorus export from an agricultural watershed: linking sources and transport mechanisms. Journal of Environmental Quality **30** 1587-1595.

Murphy, J. and H.P. Riley, 1962. A modified single solution method for the determination of phosphate in natural waters. Analytica Chimica Acta **27** 31-36.

Pionke, H., W. Gburek and A.N. Sharpley, 2000. Critical sources area controls on water quality in an agricultural watershed located in the Chesapeake Basin. Ecological Engineering **14** 325-335.

Sharpley, A.N., S. Chapra, R. Wedepohl, J.T. Sims, T. Daniel and K. Reddy, 1994. Managing agricultural phosphorus for protection of surface waters: issues and options. Journal of Environmental Quality **23** 437-451.

Sims, J., R. Simard and B. Joern, 1998. Phosphorus loss in agricultural drainage: Historical perspectives and current research. Journal of Environmental Quality **27** 277-293.

Vadas, P. and J.T. Sims, 2002. Predicting P desorption from mid-Atlantic coastal plain soils. Soil Science Society of America Journal **66** 623-631.

Effects of runoff from agricultural catchments on fishpond water chemistry: A long-term study from Třeboň fishponds

L. Pechar[1,3], J. Bastl[1], M. Hais[1], L. Kröpfelová[2], J. Pokorny[2,3], J. Stíchová[1] and J. Sulcová[2]
[1]*University of South Bohemia, Applied Ecology Laboratory, Studentská 13, 370 05 Ceské Budejovice, Czech Republic*
[2]*ENKI public benefit corporation, Dukelská 145, 379 01 Trebon, Czech Republic*
[3]*Institute of Landscape Ecology, Czech Academy of Sciences, Dukelská 145, 37901 Trebon, Czech Republic*

Abstract

Water chemistry of fishponds can be strongly subjected to direct and/or indirect effects of fishery farming and agricultural practices within a catchment. Long-term data on water chemistry of the fishponds in the Trebon Basin are related to changes in the fishery and agricultural practices, which provides an interesting tool for the description of important causal relationships. Results of the surveys in the periods 1954-1955, 1990-1991, and 2000-2001 show changes in fishpond water chemistry and trophic level. The decrease in total dissolved solids during the last ten years in fishponds reflects changes in agricultural practices within the catchment. In spite of this trend, the concentrations of nitrogen, phosphorus and chlorophyll-a remain high and they are more dependent on fishery management than on agricultural runoff.

Keywords: water chemistry, fishponds, agricultural catchments.

Introduction

The complex of fishponds in the Trebon Basin represents a unique system of artificial water reservoirs and adjacent wetlands. Fishponds represent managed aquatic ecosystems; the most important factors for their functioning (water level, fish stock, and, to some extent, also nutrients) are under human control. Since the 1930s, fish production (mainly common carp) has increased from 50 to 500 kg ha^{-1}. During the centuries, the fishponds and adjacent wetlands have become an integral part of the landscape. Presently, there are ca. 50,000 ha of fishponds in the Czech Republic, with ca. 7,000 ha in the Trebon region.

Material and methods

The Trebon region is a flat basin with an average altitude of 410-450 m above sea level. A study of the water chemistry in 35 fishponds, separated into six groups according to their catchments (Figure 1) was conducted between 1990 and 1991, and again between 2000 and 2001. Samples of water were taken three times during the vegetation growing season. At each selected site, water temperature, transparency, conductivity, pH and alkalinity were measured. Concentrations of NH_4-N, NO_3-N, dissolved reactive phosphorus (DRP), sulphates and chlorides

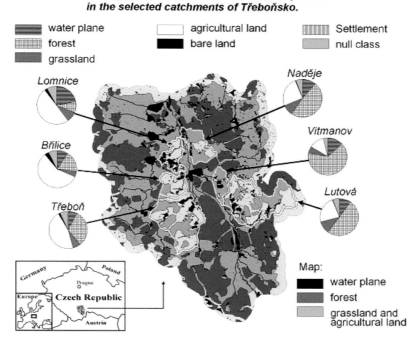

Figure 1. Land-use in the selected catchments of the Trebon basin.

(SO_4^{2-}, Cl^{-1}) were determined using a Tecator flow injection analyser. Total nitrogen (TN) and total phosphorus (TP) were determined as NO_3-N and DRP after mineralization of samples, while concentrations of the main base cations (Na^+, K^+, Ca^{2+}, Mg^{2+}) were determined by absorption spectrometry. Chlorophyll-a concentration was estimated after extraction in an acetone-methanol mixture (Pechar, 1987). Historical data were gathered from records of the Central Fishery Laboratory (Dejdar, 1954-1963); data on development of fishery management practices were provided by the Trebon Fishery Company.

Results

The management of higher fish stock densities, accompanied by higher nutrient loads, has resulted in an increase in trophic status (Figure 2). From the 1950s, mean seasonal chlorophyll-a content increased from 40 µg l^{-1} to 121 µg l^{-1}, light transparency decreased proportionally from 1.8 to 0.4 m, average concentration of total phosphorus rose from 0.16 to 0.29 mg l^{-1} and total nitrogen increased from 1.7 to 2.6 mg l^{-1} (Table 1).

Compared with data from 1954-64, results from 1990-91 show an overall increase in ionic concentrations. Total dissolved solids increased from 154 mg l^{-1} to 288 mg l^{-1}. Results from 2000-2001 show a considerable decrease of conductivity and a corresponding decrease in main ion concentrations (Table 2).

Table 1. Average trophic status of water in the fishponds at different periods between 1954 and 2001, TN = total nitrogen, TP = total phosphorus.

Years	TN (mg l^{-1})	TP (mg l^{-1})	Chlorophyll a (µg l^{-1})	Transparency (m)
1954-58	1,70	(0,16)	(40)	1,8
1990-91	2,60	0,29	121	0,45
2000-01	2,27	0,29	140	0,42

Table 2. Conductivity and concentrations of ions measured in the Trebon fishponds, 1954-1955, 1990-1991, 2000-2001.

Years	Conduct. (µS.cm^{-1})	HCO_3^- (mg l^{-1})	Cl^-	SO_4^{2-}	K^+	Na^+	Mg^{2+}	Ca^{2+}	all ion
1954-55	186	81.2	7.6	14.0	5.6	5.8	4.0	24.2	142.4
1990-91	376	125.1	24.8	69.7	11.1	11.4	8.7	36.1	286.9
2000-01	246	82.1	14.9	29.5	7.0	9.1	6.2	26.7	175.5

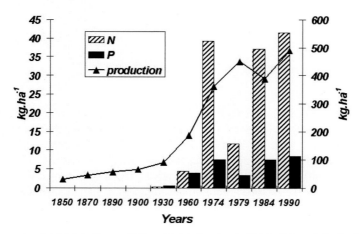

Figure 2. Average inputs of total nitrogen (TN) and total phosphorus (TP) to fishponds in the Trebon Basin between 1850 and 1990.

The most pronounced decreases in total dissolved solids were found in fishponds in the more agricultural dominated catchments. In the 10-year period, concentrations of total nitrogen and phosphorus did not change and chlorophyll-a content seemed to be slightly higher. There is no significant relationship between effect of agricultural runoff and trophic level in the fishponds, in terms of total nitrogen, phosphorus and chlorophyll-a (Figure 3a,b).

Significant relationships were found between the proportion of agricultural land in selected catchments and average values of conductivity in fishpond waters (Figure 4).

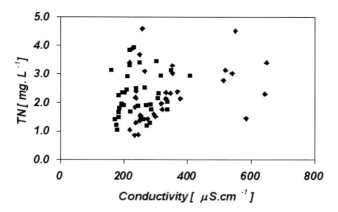

Figure 3a. Relationship between conductivity and the concentration of total N in the fish ponds of Trebon Basin (◆ data from 1990 - 1991, ■ data from 2000 - 2001).

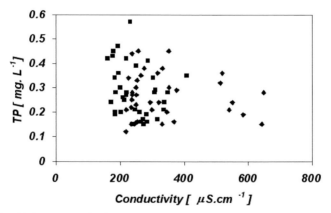

Figure 3b. Relationship between conductivity and the concentration of total P in the fish ponds of Trebon Basin (◆ data from 1990 - 1991, ■ data from 2000 - 2001).

Discussion

From the 1950s to the end of the 1980s, the concentrations of ions and compounds of nitrogen and phosphorus increased considerably in fishpond waters. During the 1950s, fishponds were limed and heavily fertilised with superphosphate and ammonium nitrate. From the 1960s to mid-1980s, the use of mineral fertilizers was replaced by applications of manure. The long-term input of nutrients (N and P) is reflected in the high concentrations of these nutrients in waters, as well as in the increased phytoplankton biomass. The results of these changes are large fluctuations in pH and dissolved oxygen. The overloading of fishponds by manure inputs, results (during the summer months) in increased available phosphorus, but also nitrogen deficiency (Pechar *et al.*, 2002) as addition of organic fertilizer enhances decomposition processes. Decreases in oxygen content in surface sediment leads to intense denitrification.

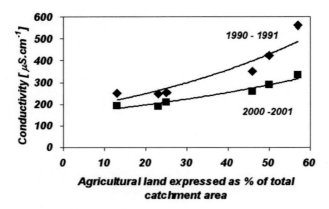

Figure 4. Relationships between the proportion of agricultural land in catchments and average values of conductivity.

The development of phytoplankton causes a rise of pH above 9 whereby free ammonia is volatilised (Pokorny *et al.*, 1999).

Unlike fishery management, agriculture has been subjected to radical changes occurring since 1989. Before 1989, nearly all agricultural land was used by state-owned enterprises or co-operative farms, and intensive agriculture and production of pigs and cattle were subsidised by state agencies. Since 1991, land ownership has radically changed and the whole structure of agricultural farms has also changed (Jeník *et al.*, 2002). The reduced eutrophication of surface waters, due to lower applications of fertilizers and smaller production of manure, has resulted in significant decreases in agricultural runoff. In spite of the decreasing trend in the input of main ions from the catchment, the overall trophic level of Trebon fishponds still remains very high.

Conclusion

The transformation in the Czech Republic since 1989 initiated rapid changes in land ownership and in agricultural practices. Results from the years 2000-2001 show a considerable decrease in main ion concentrations in fishpond waters. In the ten years evaluated, the concentrations of total nitrogen and phosphorus did not change, while chlorophyll-a content seems higher. There is no significant relationship between the effect of agricultural runoff and trophic levels of fishponds in terms of total nitrogen, phosphorus and chlorophyll a. Trophic level depends on fishery management such as fish stock and amount of directly-applied fertilizers.

Acknowledgements

This study was supported by the projects MSM 122200003, MSM 6007665806 and MSM 000020001 of the Ministry of Education of the Czech Republic.

References

Dejdar, K., 1954-1963. Unpublished data., Archives of the Institute of Botany, Czech Academy of Sciences, Trebon.

Jeník, J., M. Hátle, and J. Hlásek, 2002. Preservation of ecological and socio-economic roles of human-managed wetlands. In: Freshwater Wetlands and Their Sustainable Future: A Case Study of the Trebon Basin Biosphere Reserve, Czech Republic. Man and the Biosphere Series **28**, edited by J. Kvet J., J. Jeník, and L. Soukupová, UNESCO and The Parthenon Paris, pp. 481 - 486.

Pechar, L., 1987. Use of the acetone-methanol mixture for extraction and spectrophotometric determination of chlorophyll a in phytoplankton. Arch. Hydrobiol. Suppl. **78** 99-117.

Pechar, L., I. Prikryl, and R. Faina, 2002. Hydrobiological evaluation of Trebon fishponds since the end of 19th century. In: Freshwater Wetlands and Their Sustainable Future: A Case Study of the Trebon Basin Biosphere Reserve, Czech Republic. Man and the Biosphere Series **28**, edited by J. Kvet J., J. Jeník, and L. Soukupová, UNESCO & The Parthenon Paris, pp. 31-62.

Pokorny, J., S. Fleischer, L. Pechar, and J. Pansar, 1999. Nitrogen distribution in hypertrophic fishponds and composition of gas produced in sediment. In: Nutrient Cycling and Retention in Natural and Constructed Wetlands, edited J. Vymazal, Backhuys Publishers, Leiden, The Netherlands, pp.111-120.

Assessment of landscape efficiency in matter retention in submontane agricultural catchments of the Czech Republic

L. Pechar[1,3], J. Procházka[1], M. Hais[1], E. Pecharová[1], M. Eiseltová[2], L. Bodlák[2], J. Sulcová[2] and L. Kröpfelová[2]
[1]University of South Bohemia, Applied Ecology Laboratory, Studentská 13, 370 05 Ceské Budejovice, Czech Republic
[2]ENKI public benefit corporation, Dukelská 145, 379 01 Trebon, Czech Republic
[3]Institute of Landscape Ecology, Czech Academy of Sciences, Dukelská 145, 37901 Trebon, Czech Republic

Abstract

The evaluation of landscape functions is a crucial problem of landscape management and an important aspect of ecological research. A new ecological concept of landscape efficiency has been used to study the processes and system functioning within selected submontane agricultural catchments. The concept of landscape efficiency measures the interrelationship of energy dissipation with water and matter flows. The characteristic features of good landscape efficiency and sustainability are: closed matter cycling, balanced water budget and low matter losses with water discharges. Catchments with implemented drainage and predominately arable farming showed a lower performance of landscape functions, for example, reduced water-retention capacity and quality of surface water discharged.

Keywords: landscape sustainability, matter losses, temperature distribution, water discharge.

Introduction

Today's unsustainable management of natural resources, such as soil, water and vegetation, results in a serious degradation of ecological systems and the landscape as a whole. When not disturbed, a landscape will tend to enhance and/or maintain its sustainability. Nature operates in cycles maintaining short-circuited cycles of elements, such as nutrients, minerals and water. The fact that vegetation plays an important role has been shown by studies of post-glacial development of southern Swedish lakes, which revealed a clear interdependence of catchment vegetation cover and the deposition rates of sediments including nutrients and base cations (Digerfeldt, 1972). High sedimentation rates, which occurred after the ice sheet retreated enabled vegetation to develop over the entire catchment. This strongly infers that abundant vegetation controls water and matter cycles. Supported by these findings, landscape sustainability has been defined as the efficiency of the landscape to recycle water and matter and to dissipate the solar energy pulse in time and space (Ripl, 1995; Ripl *et al.*, 1995). Our study was aimed to test these methods for evaluation of matter losses and thermal distribution and to verify the relevance of this approach in small catchments.

Materials and methods

Site description

Two submontane areas, Zalipno and the Upper Stropnice River catchment are located in southern Bohemia (Czech Republic). In Zalipno, the three small catchments studied are of similar geographical conditions but presently under different land uses (Table 1). In the Upper Stropnice catchment, two small catchments differing in land use were selected. Pasecky stream catchment has no arable land and a large proportion of forest, whilst Váckovy stream catchment has a high proportion of arable land (Table 2).

Table 1. Basic characteristics of the catchments in Zalipno.

	Mlynsky	Horsky	Bukovy
Catchment area (ha)	214,1	201,7	264,4
Altitude (m above sea level)	784 - 884	826 - 1026	809 - 1026
non-forested (ha)	195,6	56,4	12,8
Management of non-forested areas	pasture, mowed meadows	non-manged areas, mowed meadows	mowed meadows

Table 2. Land use of the catchments in Upper Stropnice River region.

Catchment	Total area [km²]	Forest [km²]	Arable land [km²]	Grassland [km²]	Settlements [km²]
Pasecky	2,62	1,83	0	0,73	0,06
Váckovy	2,34	1,32	0,98	0,04	0

Precipitation, waterflow, matter losses and thermal distribution have been systematically monitored since 1998 in Zalipno, and since 2000 in the Upper Stropnice catchment. The catchments were equipped with an automatic monitoring station measuring at 20-minute intervals waterflow, conductivity and temperature of the water. Chemical analyses of waters included pH, alkalinity and concentrations of main ions. The concentrations of NH_4-N, NO_2-N, NO_3-N, PO_4-P, Cl^-, and SO_4^{2-}, Ca^{2+}, Mg^{2+}, K^+, Na^+ were measured (Procházka *et al.*, 2001). The distribution of temperature was determined from a Landsat TM 5 satellite picture of 27 July 2003.

Results

In Zalipno, the highest rate of ion losses was found in the drained Mlynsky catchment (Table 3). The chemical composition of water leaving the Mlynsky catchment indicates rapid mineralization of soil organic matter. A similar situation was found in the two (which were distinctly different in land use) sub-catchments within the Upper Stropnice catchment. Conductivity and leaching of main ions are much higher in the intensively farmed Váckovy catchment compared to the Pasecky catchment, which is mainly covered by forest and grassland (Table 4). Distinctive differences in waterflow patterns have been observed between the drained, intensively-used, and those catchments with more natural vegetation cover and without intensive drainage. The waterflow in Mlynsky catchment fluctuates markedly compared to both the Bukovy and Horsky catchments, where water dicharge is much more even (Figure 1); this reflects poor water retention in the drained catchment.

Table 3. Conductivity (μS cm^{-1}) and concentration of main ions (mg l^{-1}) in discharge water from the catchments in Zalipno.

	Mlynsky			Horsky			Bukovy			
	n	x	±SD	n	x	±SD	n	x	±SD	p level
conductivity	55	93.07	18.2 a	58	45.22	15.7 b	58	36.07	11.1 c	0.000
pH	62	6.37	0.3 a	62	6.07	0.5 b	62	5.97	0.6 c	0.000
HCO_3^-	61	0.46	0.1 a	60	0.22	0.1 b	58	0.16	0.1 a	0.000
NO_3^-	63	8.15	2.5 a	63	2.12	2.3 b	63	2.41	6.6 a	0.000
Cl^-	63	1.77	1.3 a	63	1.07	0.8 b	63	1.08	1.8 a	0.004
SO_4^{2-}	62	12.73	6.8	61	11.56	6.0	61	11.52	8.2	0.582
Ca^{2+}	55	7.65	2.0 a	55	3.10	1.2 b	56	2.83	4.3 a	0.000
Mg^{2+}	55	1.62	1.0 a	55	0.80	0.4 b	56	0.63	0.4 a	0.000
Na^+	55	3.66	0.8 a	55	2.50	0.8 b	56	2.63	0.8 a	0.000
K^+	55	1.58	0.3 a	55	1.12	0.3 b	56	0.53	0.3 c	0.000

x -mean, SD -standard deviation, n -number of samples; different letters in the rows indicate significant differences.

Table 4. Conductivity (μS cm^{-1}) and concentration of main ions (mg l^{-1}) in discharge water from the catchments in Upper Stropnice River region.

	Cond.	HCO_3^-	NO_3^-	NO_2^-	NH_4^+	PO_4^{3-}	Cl^-	SO_4^{2-}	Ca^{2+}	Mg^{2+}	Na^+	K^+
Pasecky	81	19.15	5.20	0.01	0.21	0.05	2.97	14.60	6.60	1.85	4.40	1.89
Váckov	219	33.84	16.73	0.03	0.05	0.04	6.56	69.32	20.32	6.35	6.83	2.15

Landsat 5 TM satellite images (using channel 6 -temperature) show the differences in surface temperatures in three studied catchments in Zalipno region. Drained Mlynsky catchment differs from the other two catchments, as well as from the surrounding Danube Basin reference area, by its heated surfaces (Figure 2).

Conclusions

Our endeavour to remove water from landscapes such as those illustrated in this study by the implementation of systematic drainage networks on arable land, surface drains in forests, straightening of streams, drainage of floodplains along with destruction of natural vegetation

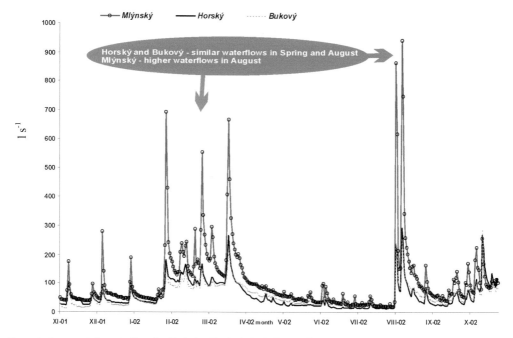

Figure 1. Average daily waterflow (l s^{-1}) during the hydrological year of 2002.

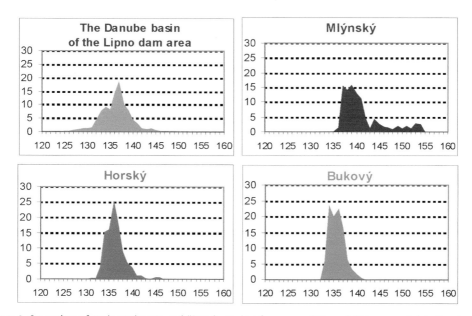

Figure 2. Comparison of study catchments and "Danube Basin reference area" through histograms of data from Landsat 5 TM-6 band. The x-axes are the radiometric pixel values (high value = high relative temperature) and the y-axes are the ratio of overall pixel number

cover over large areas, has resulted in increased irreversible losses of life-supporting elements (nutrients and base cations) and highly-fluctuating discharges in streams (causing both flooding and droughts). This study has shown that excessive transport of nutrients such as nitrogen and phosphorus goes hand in hand with the changes in the water cycle and corresponding changes in temperature distribution over the landscape.

Non-point source pollution is responsible for a significant part of water pollution. Therefore, the attention of society urgently needs to focus on improving management practices over entire landscapes. Agricultural policies need to provide incentives to land managers for the implementation of more sustainable land management practices. The presented monitoring approach could be used to identify the most sensitive areas that require more sustainable land management.

Acknowledgements

This study was supported by the projects MSM 122200003, MSM 6007665806 and MSM 000020001 of the Ministry of Education of the Czech Republic.

References

Digerfeldt, G., 1972. The Post-Glacial Development of Lake Trummen. Regional vegetation history, water level changes and palaeolimnology. Folia Limnologica Scandinavica **16**, 104 pp.

Procházka, J., P. Hakrová, J. Pokorny, E. Pecharová, T. Hezina, M. Síma, and L. Pechar, 2001. Effect of different management practices on vegetation development, losses of soluble matter and solar energy dissipation in three small sub-mountain catchments. In: Transformations of Nutrients in Natural and Constructed Wetlands, edited by J. Vymazal, Backhuys Publishers, Leiden, The Netherlands, pp. 143-175.

Ripl, W., 1995. Management of water cycle and energy flow for ecosystem control: the energy-transport-reaction (ETR) model. Ecological Modelling **78** 61-76.

Ripl, W., C. Hildmann, T. Janssen, I. Gerlach, S. Heller, and S. Ridgill, 1995. Sustainable redevelopment of a river and its catchment: the Stör River project. In: Restoration of Stream Ecosystems - an integrated catchment approach, edited by M. Eiseltová and J. Biggs. IWRB Publ. **37**, pp. 76-112.

Wetlands of Ireland - An overview: diversity, ecosystem services and utilization

M.L. Otte
Wetland Ecology Research Group, Department of Botany, University College Dublin, Belfield, Dublin 4, Ireland

Abstract

Ireland is rich in wetlands, which provide many invaluable ecosystem services. Natural wetlands can be used as models for constructed wetlands for quality improvement of wastewater and other applications. In Ireland, the value and potential for utilization of natural and artificial wetlands is only now being realized. This paper gives a brief overview of the wetlands occurring naturally in Ireland and then reviews some aspects relevant to constructed wetlands for wastewater treatment in an Irish context.

Keywords: biodiversity, ecology, functions, economic value.

Introduction

Due to its Atlantic climate, with a precipitation rate exceeding the rate of evapotranspiration throughout the year, Ireland is rich in water. In fact it ranks as the eighth water-richest country in the world according to the Water Poverty Index, which is currently being developed at the Centre for Ecology and Hydrology in Wallingford, UK (Sullivan *et al.*, 2003, see also http://www.nerc-wallingford.ac.uk/research/WPI/). It is therefore not surprising that Ireland is rich in wetlands, not just bogs for which it is perhaps best known (Doyle 1990, Feehan and O'Donovan, 1996), but also salt marshes, dune slacks, swamps, fens, floodplains, lagoons, wet woodlands, and turloughs (Otte, 2003). The latter refers to ephemeral lakes (Irish, 'tuar loch' meaning 'dry lake'), located mainly in the karstic region on the border of counties Clare and Galway, and of which very few exist outside Ireland (Goodwillie and Reynolds, 2003).

Wetlands are generally defined (Mitsch and Gosselink, 2000) as areas that are inundated or saturated by surface or groundwater at a frequency sufficient to support a prevalence of vegetation typically adapted for life in saturated soil conditions. This definition, which has been the basis for legislation pertaining to protection of wetlands in the United States of America for several decades, thus combines three characteristics: variable hydrology, hydric soils, and vegetation consisting of emergent plants (helophytes). A somewhat wider definition for wetlands was agreed in the Convention on Wetlands, also known as the Ramsar Convention. This is an intergovernmental treaty signed in Ramsar, Iran, in 1971, which provides the framework for national action and international cooperation for the conservation and wise use of wetlands and their resources. It defined wetlands as areas of marsh, fen, peatland or water, whether natural or artificial, permanent or temporary, with water that is static, flowing, fresh, brackish or salt, including areas of marine water, the depth of which at low tide does not exceed six meters (see www.ramsar.org). The latter definition therefore includes more open water.

Wetlands are often ecotones, ecosystems that form a transitional gradient between two extremes, in this case open water and dry land, having characteristics of both.

Ecosystem functions, ecosystem services and the value of wetlands

Ecosystem functions are processes carried out by the whole ecosystem, rather than its individual components. Such functions include biomass production, cycling of nutrients and capture of the sun's energy through photosynthesis. The term 'ecosystem services' (Cairns 1993, Daily *et al*. 1997) refers to the same processes, but emphasizes their importance to the environment as a whole and to humans in particular (Table 1).

This can be used to calculate a monetary value for the services provided by wetlands and other ecosystems (Constanza *et al*., 1997; Mitsch and Gosselink, 2000). For example, what would it cost to remove pollutants from wastewater by mechanical and chemical means as effectively as it is done by wetlands, or to convert energy from the sun into food for animals? Looking at it this way, it becomes clear that wetlands provide invaluable ecosystem services, and that

Table 1. Ecosystem functions, or services, of wetlands (Mitsch and Gosselink, 2000; Otte, 2003).

Function	Description
● Production of biomass	Wetlands are among the most productive ecosystems in the world, effectively turning energy from sunlight into the materials that drive life. Through this process also, wetlands provide fuel in the form of peat and other 'biofuels', such as wood from coppice willow (Van den Broek *et al*. 2002), and building materials, such as reeds for thatched roofs.
● Cycling of carbon, and nitrogen and other plant nutrients	Due to their high productivity, wetlands sequester carbon and plant nutrients, which are very effectively cycled through soil, water, biomass and the atmosphere. As such, wetlands form a significant component of global elemental cycles.
● Hydraulic moderators	Wetlands act as buffers against floods and also retain water during dry spells.
● Improvement of water quality	Many pollutants - nitrogen, phosphate, metals, organic pollutants such as pesticides, and biological pollutants such as coliform bacteria - are very effectively filtered from water due to the physical and biogeochemical conditions of the sediments and soils, and due to uptake by plants.
● 'Highways' for wildlife	In urbanized areas particularly, wetlands are the only pathways linking ecosystems, such as lakes, woodlands and open fields, and thus provide migratory routes for wildlife.
● Shelter and food for wildlife	Many animals depend on wetlands for shelter and food, including invertebrates, birds, fish, amphibians, and mammals (Hutchinson 1979, Sheppard 1993, Reynolds 1998, Healy 2003, Nairn 2003, Reynolds 2003).
● Hunting and fisheries	Their richness in animal life traditionally made wetlands valuable as hunting grounds. For many fish, wetlands are important breeding grounds, and thus of invaluable importance to the fishing industry.
● Aesthetic value and recreation	Many wetlands are simply beautiful places which form integral parts of the Irish landscape

there is as yet no technical alternative for several of those (e.g. turning sunlight into food for animals).

Putting a monetary value on wetlands can be used for raising awareness of wetland related issues. Unfortunately perhaps, a statement that a particular wetland must be protected because its ecosystem services are valued at, say, €10,000 ha^{-1} yr^{-1} is usually more convincing than to argue that it is valuable because the very rare grass *Puccinellia fasciculata* occurs there. Putting a monetary value on wetlands can also aid in management decisions. For example, would wastewater treatment by the traditional chemical/mechanical means be more or less cost-effective than a wetland constructed to carry out the same task?

Wetland valuation, however, is not an easy task. While some ecosystem services are relatively easily expressed in terms of monetary value, others are not. For example, the value of coastal wetlands to the fishing industry has been studied in some detail in several countries and can be expressed as a value per hectare (100x100 m^2, ha) per year. Turner (1991) calculated a value for coastal wetlands in Thailand to the shrimp fishing industry of US$160-500 ha^{-1} yr^{-1}. However, how should the capacity of wetlands to capture sunlight and convert it to biomass be valued? Similarly, how would one value the quality improvement capacity of natural wetlands for surface water, or their value as habitats for flora and fauna? Still, attempts have been made, and the analysis clearly shows that in terms of monetary value for ecosystem services, wetlands rank very high indeed. Constanza *et al.* (1997), going by 1994 rates, calculated average global monetary values of ecosystem services equivalent to US$ 9,990-22,832 ha^{-1} yr^{-1} for wetlands. This compared to US$ 969 ha^{-1} yr^{-1} for forests, US$ 232 ha^{-1} yr^{-1} for grasslands and only US$ 92 for croplands (of course, excluding the value of the crops themselves, because that is an economic value to the farmer, instead of the value of a service to the environment and human society).

Constructed wetlands and other applications of wetlands

With the realization that wetlands supply many services, interest in utilizing these services grew. Not only did it become clear that natural wetlands were not wastelands and deserved protection, but also that wetlands could be constructed to carry out specific functions, such as quality improvement of wastewater. In this respect Ireland has been well behind other countries. While in the USA policy in relation to protection of wetlands goes as far back as 1789 (Tzoumis, 1998) and constructed wetlands have been in use for many decades (Kadlec and Knight, 1996; Mitsch and Gosselink, 2000), such systems in Ireland have not had the support from local or national governments and implementation has therefore been slow and often haphazard (Harty and Otte, 2003). Owing to the lack of formal support, wetlands constructed for wastewater treatment were not widely advertised and the valuable information that could have been obtained from these pioneering systems in Ireland were lost to the scientific and general community. It should therefore not surprise that the landmark publication on constructed wetlands in Europe by Vymazal *et al.* (1998) includes individual chapters for most European countries with specific information on their constructed wetlands, but Ireland is not included. Yet, a wide range in terms of design and applications of between 100 and 150 constructed wetlands now exists in Ireland (Harty and Otte, 2003, http://www.ucd.ie/wetland/construcwet/constructedwetlands.htm). Municipal systems treat wastewater of single households and small communities, campgrounds, holiday villages, schools and visitor centres. Several farms around the country have constructed wetlands to treat their agricultural wastewater, while industrial applications include the dairy industry, a meat packaging plant and a system treating mine wastewater rich in sulfate at Tara Mines, Navan.

Wetlands can be constructed for many reasons, not just for wastewater treatment. For example, floodplain wetlands along the Rhine river which had been destroyed in the past are now being rehabilitated in order to restore the biodiversity and the hydraulic buffering capacity of the river (Nienhuis *et al.*, 2002). Another application is the creation of artificial wetlands on mine tailings (phytostabilization), which immobilizes the potentially toxic components of the mine tailings and prevents dust blows (McCabe, 1998; McCabe and Otte, 2000). While there is no official record of the loss of wetlands in Ireland it is obvious that a very large area has been destroyed due to arterial drainage schemes and peat extraction (e.g. see Doyle and Críodáin, 2003). Yet rehabilitation of Irish wetlands is only now being considered and very few scientific reports (e.g. Holden *et al.*, 2004) have been published.

Whatever the reasons for constructing a wetland, it must be realized that artificial wetlands can not replace natural wetlands (see for example Zedler 2000). Constructed wetlands will differ in soil composition and, particularly if they are relatively young, have a lower biodiversity than natural wetlands. Because the efficiency of ecosystem services typically increases with increasing biodiversity (Naeem *et al.*, 1994; Zedler *et al.*, 2001; Callaway *et al.*, 2003) constructed wetlands tend to be less efficient and stable compared to natural wetlands. From this point of view it is striking that despite the wide variation in design and pollutants to be removed, very little variation exists in terms of planting of constructed wetlands. Most systems, in Ireland and abroad, are planted predominantly with *Phragmites australis* (reeds) and/or *Typha* spp. (cattails, bullrush). Considering the wide range of types of wetlands in Ireland alone, there is huge potential for further development of constructed wetlands. At the same time, care should be taken with the choice of plant species to be used in constructed wetlands. Biodiversity does not mean as many species as possible regardless of origin. Biodiversity means diversity in species and ecotypes appropriate to the ecosystems occurring naturally in a specific area. One reason is that biodiversity is not just about species numbers, but also about retaining genetic variation, thus ensuring the stability of communities under varying environmental conditions. Another reason is that non-native plants can become invasive and thus eventually reduce biodiversity by out-competing native species (Cronk and Fuller, 1995; Child *et al.*, 2003, see also the WebPages of the Center for Aquatic and Invasive Plants at http://plants.ifas.ufl.edu/). It would therefore be desirable to design constructed wetlands using local natural wetlands as models, with plants sourced locally as well.

Longevity of constructed wetlands

The longevity of constructed wetlands is of concern (Walski, 1993; Drizo *et al.*, 2002), particularly to licensing authorities such as the Environmental Protection Agency of Ireland. It is however difficult to accurately predict longevity because most systems are no older than a few decades. Perhaps we can learn from natural wetlands. In reply to a question by Walski (1993) on this issue in relation to retention of metals, Hedin (1996) and Williams and Stark (1996) based on their studies of constructed wetlands, estimated their longevity to be in the order of a few decades only. However, Beining and Otte (1997), investigating metal retention in a natural, so called 'volunteer' wetland in Glendalough, Co. Wicklow, Ireland, clearly demonstrated that that system had been effectively removing metals from mine run-off for at least a century. They further calculated that it had the potential to continue to do so for at least another two centuries. As long as a wetland is large enough relative to loading rates its life-expectancy should be sufficient to make constructed wetlands viable long-term alternatives for traditional methods of wastewater treatment. Besides, when the end of its life span is reached, the pollutants retained in wetland sediments and vegetation may well be recycled. For example, plant nutrients, particularly phosphate, are valuable as fertilizer and could be re-used. Similarly, more persistent pollutants, such as metals, in treatment wetlands are effectively returned to the form from which they were originally derived, bound as precipitates, and can thus be recycled as well.

Conclusion

The attitudes towards wetlands in Ireland are slowly changing. The traditionally agriculture based Irish community long viewed wetlands as worthless land, that could at best be drained or filled in. Now it is gradually being realized that wetlands provide services that are invaluable, such as retention of pollutants and flood control. While so far no study has attempted to put a monetary value on Irish wetlands, it is clear from studies abroad that their value is likely to be many times that of the crops that may be commercially produced on them after drainage. This raises the question whether or not wetland ecosystems should be specifically protected by law. This is the case in the USA, where wetlands have been protected by Federal and State laws since the mid-1970s (Mitsch and Gosselink, 2000). However, while the US-approach can be used as a model, it can not simply be applied to Ireland. For example, to be protected a clear definition of what are wetlands must be agreed on, but if the definitions mentioned above (see Introduction) would be used for legal protection in Ireland, few areas on the island would not be wetland. Such an approach would not be helpful, and a more pragmatic solution is therefore to rank wetlands according to their importance, assessed not just by their rarity as ecosystems or the occurrence of rare species, but also by their function as an integral part of the landscape, as has been proposed by the Heritage Council (2002). For an in-depth review of the current state of affairs concerning conservation and management of wetlands in Ireland see Clabby (2003).

While the destruction of natural wetlands in Ireland needs to be halted, there is also huge potential for constructing wetlands for improvement of quality of wastewater, including that arising from diffuse sources such as agricultural run-off and from point sources such as farm-yards. Wetland restoration and construction should be carried out with care and based on sound science. It is not just a matter of digging a hole and planting a few reeds. And size matters - constructed wetlands have to be large enough relative to the loading rates, and so may not be feasible in every location.

Constructed wetlands in Ireland have had no lack in enthusiastic supporters, but their development has been strongly hampered by opposition from government agencies at the local and national levels, despite ample proof from abroad about their efficiency and reliability. This attitude can partly be blamed on a lack of understanding of wetlands in general. It is time to change, and the signs are that this is finally happening. Just as the Swiss appreciate their mountains, Ireland should embrace its wetlands.

Acknowledgements

Donna Jacob for proof-reading the manuscript. All past and present members of the Wetland Ecology Research Group, Department of Botany, UCD, without whom, even less would be known about Irish wetlands.

References

Beining, B.A. and M.L. Otte, 1997. Retention of metals and longevity of a wetland receiving mine leachate. In: Proceedings of the 14th Annual National Meeting - Vision 2000: An Environmental Commitment. American Society for Surface Mining and Reclamation, Austin, Texas, May 10 - 16, 1997. edited by J.E. Brandt, J.R. Galevotic, L. Kost, and J. Trouart, ASMR, Lexington, KY, USA, pp. 43 - 46.

Cairns, J. Jr. 1993. Determining desirable ecosystem services per capita. Journal of Aquatic Ecosystem Health **2** 237-242.

Callaway, J.C., G. Sullivan, and J.B. Zedler, 2003. Species-rich plantings increase biomass and nitrogen accumulation in a wetland restoration experiment. Ecological Applications **13** 1626-1639.

Child, L., J.H. Brock, G. Brundu, K. Prach, P. Pysek, P.M. Wad,e and M. Williamson, 2003, Eds. Plant invasions - Ecological threats and management solutions. Backhuys Publishers, Leiden, The Netherlands.

Clabby, G. 2003. Conservation and management of wetlands in Ireland. In: Wetlands of Ireland - Distribution, Ecology, Uses and Economic Value, edited by Otte, M.L., UCD Press, Dublin, pp. 213-218.

Constanza, R., R. d'Arge, R. de Groot, S. Farber, M. Grasso, B. Hannon, K. Limburg, S. Naeem, R.V. O'Neill, J. Paruelo, R.G. Raskin, P. Sutton, and M. van den Belt, 1997. The value of the world's ecosystem services and natural capital. Nature **387** 253-260.

Cronk, Q.C.B. and J.L. Fuller, 1995. Plant invaders. Chapman & Hall, London, 241 pp.

Daily, G.C., S. Alexander, P.R. Ehrlich, L. Goulder, J. Lubchenco, P.A. Matson, H.A. Mooney, S. Postel, S.H. Schneider, D. Tilman, and G.M. Woodwell, 1997. Ecosystem services: benefits supplied to human societies by natural ecosystems. Issues in Ecology 2. Ecological Society of America, Washington, D.C., 16 pp.

Doyle, G.J. 1990. Ecology and conservation of Irish peatlands. Royal Irish Academy, Dublin, 221 pp.

Doyle, G.J. and C. Críodáin 2003. Peatlands - fens and bogs. In: Wetlands of Ireland - Distribution, Ecology, Uses and Economic Value, edited by Otte, M.L., UCD Press, Dublin, pp. 79-108.

Drizo, A., Y. Comeau C. Forget and R.P. Chapuis 2002. Phosphorus saturation potential: a parameter for estimating the longevity of constructed wetland systems. Envirnmental Science and Technology **36** 4642-4648.

Feehan, J. and G. O'Donovan, 1996. The bogs of Ireland: an introduction to the natural, cultural and industrial heritage of Irish peatlands. Environmental Institute, University College Dublin, Dublin, 518 pp.

Goodwillie, R. and J.D. Reynolds, 2003. Turloughs. In: Wetlands of Ireland - Distribution, Ecology, Uses and Economic Value, edited by Otte, M.L., UCD Press, Dublin, pp. 130-134.

Harty, F. and M.L. Otte, 2003. Constructed wetlands for treatment of wastewater. In: Wetlands of Ireland - Distribution, Ecology, Uses and Economic Value, edited by Otte, M.L., UCD Press, Dublin, pp. 182-190.

Healy, B. 2003. Coastal lagoons. In: Wetlands of Ireland - Distribution, Ecology, Uses and Economic Value, edited by Otte, M.L., UCD Press, Dublin, pp. 51-78.

Hedin, R.S. 1996. Environmental engineering forum: Long-term effects of wetland treatment of mine drainage. Journal of Environmental Engineering **122** 83-84.

Heritage Council. Integrating policies for Ireland's landscape. The Heritage Council, Kilkenny.

Holden, J., P.J. Chapman, and J.C. Labadz, 2004. Artificial drainage of peatlands: hydrological and hydrochemical process and wetland restoration. Progress in physical geography **28** 95-123.

Hutchinson, C.D. 1970. Ireland's wetlands and their birds. Irish Wildlife Conservancy, Dublin, 201 pp.

Kadlec, R.H. and R.L. Knight, 1996. Treatment wetlands. Lewis publishers. Boca Raton, 893 pp.

McCabe, O.M. 1998. Wetland plants for revegetation of metal mine tailings. PhD thesis, University College Dublin, Dublin, 257 pp.

McCabe, O.M. and M.L. Otte, 2000. The wetland grass *Glyceria fluitans* for revegetation of mine tailings. Wetlands **20** 548-559.

Mitsch, W.J. and J.G. Gosselink, 2000. Wetlands. 3rd Edn. John Wiley & Sons, New York, 750 pp.

Naeem, S., L.J. Thompson, S.P. Lawler, J.H. Lawton, and R.M. Woodfin, 1994. Declining biodiversity can alter the performance of ecosystems. Nature **368** 734-737.

Nairn, R.W. 2003. Birds of Irish wetlands: a literature review. In: Wetlands of Ireland - Distribution, Ecology, Uses and Economic Value, edited by Otte, M.L., UCD Press, Dublin, pp.197-201.

Nienhuis, P.H., A.D. Buijse, R.S.E.W. Leuven, A.J.M. Smits, R.J.W. de Nooij, and E.M. Samborska, 2002. Ecological rehabilitation of the lowland basin of the river Rhine NW Europe. Hydrobiologia **478** 53-72.

Otte, M.L. 2003, Ed. Wetlands of Ireland - Distribution, Ecology, Uses and Economic Value. UCD Press, Dublin, 256 pp.

Reynolds, J.D. 1998. Ireland's freshwaters. The Marine Institute, Dublin, 130 pp.

Reynolds, J.D. 2003. Fauna of turloughs and other wetlands. In: Wetlands of Ireland - Distribution, Ecology, Uses and Economic Value, edited by Otte, M.L., UCD Press, Dublin, pp. 145-156.

Sheppard, R. 1993. Ireland's wetland wealth. Irish Wildlife Conservancy, Dublin, 152 pp.

Sullivan, C.A., J.R. Meigh, A.M. Giacomello, T. Fediw, P. Lawrence, M. Samad, S. Mlote, C. Hutton, J.A. Allan, R.E. Schulze, D.J.M. Dlamini, W. Cosgrove, J.D. Priscoli, P. Gleick, I. Smout, J. Cobbing, R. Calow, C. Hunt, A. Hussain, M.C. Acreman, J. King, S. Malomo, E.L. Tate, D. O'Regan, S. Milner and I. Steyl, 2003. The water poverty index: Development and application at the community scale Natural Resources Forum **27** 189-199.

Turner, K. 1991. Economics and wetland management. Ambio **20** 59-63.

Tzoumis, K.A. 1998. Wetland policymaking in the US Congress from 1789 to 1995. Wetlands **18** 447-459.

Van den Broek, R., A. Van Wijk, and W. Turkenburg, 2002. Electricity from energy crops in different settings - a country comparison between Nicaragua, Ireland and The Netherlands. Biomass and Bioenergy **22** 79-98.

Vymazal, J., H. Brix, P.F. Cooper, M.B. Green, and R. Haberl, 1998. Constructed wetlands for wastewater treatment in Europe. Backhuys Publishers, Leiden, The Netherlands, 366 pp.

Walski, T.M. 1993. Long-term effects of wetland treatment of mine drainage. Journal of Environmental Engineering **119** 1004-1005.

Williams, F.M. and L.R. Stark, 1996. Environmental engineering forum: Long-term effects of wetland treatment of mine drainage - Managed wetlands may be long-term solutions to mine-water treatment. Journal of Environmental Engineering **122** 84-85.

Zedler, J.B. 2000. Progress in wetland restoration ecology. Trends in ecology and evolution **15** 402-407.

Zedler, J.B., J.C. Callaway, and G. Sullivan, 2001. Declining biodiversity: Why species matter and how their functions might be restored in Californian tidal marshes. Bioscience **51** 1005-1017.

Wetland functions and values in agricultural watersheds: an American perspective

M.W. Clark
Wetland Biogeochemistry Laboratory, Soil and Water Science Department, University of Florida/IFAS, 106 Newell Hall, PO Box 110510, Gainesville, FL 32611-0510, USA

Introduction

In the United States wetlands are a major focal point for environmental policy issues and a president setting arena in the balance between private property rights and public benefits (Kohn, 1994; Holtman *et al.*, 1996). As a society, our understanding of the role and function of wetlands has increased considerably during the past 50 years. As a result, our regulatory policies have also changed, now affording wetlands some of the strongest regulatory protection measures of any ecosystem in the landscape. In agricultural areas the conversion of wetlands to crop or grazing lands seems the only profitable land use in an industry where intensification and optimisation are often a necessity. However, in light of increasing regulatory mandates to maintain water quality, wetlands are being viewed in a new paradigm, one that may benefit both private agricultural interests, while preserving the broader societal values of clean water, wildlife habitat and ecological functions; however this was not always the case.

Period of wetland exploitation

In the lower 48 contiguous states, at the time when the 13 original colonies were becoming established, wetlands covered an estimated 89.4-90.6 million hectares of the landscape (Dahl, 1990). The highest percentage of wetland area occurred in what would become states along the Eastern seaboard, Midwestern region around the Great Lakes, upper Mississippi River watershed, and the Gulf coast states where many broad floodplains and delta's exist. These regions are lower in topographic relief, and rainfall in most years is equal to or greater than evapotranspiration. Soils associated with these wetland communities are often rich in organic matter with high fertility due to alluvial deposits from upland erosion. Although not the first choice for agriculture, at a time when irrigation technology was primitive and inorganic fertilizer use was either non existent or cost prohibitive, these low lying areas with available soil moisture and high soil fertility were attractive dependent on flooding frequency and/or excessive soil moisture conditions.

Prior to the later part of the 19th century, technological advances limited the rate at which wetland conversion occurred in the United States (Heimlich *et al.*, 1998). However, even during the early part of the 19th century, significant drainage of wetlands by damming and diking along rivers, and drainage of bottomlands began to convert wetlands for agriculture and other land use activities. Government supported incentives were also instrumental in transferring public land to states and eventually to private landowners to promote agriculture. The Swamplands Act of 1849, 1850 and 1860 gave states a total of 26.3 million hectares of government owned wetlands for agricultural "reclamation" (Shaw and Fredine, 1971). In the Midwestern states this was principally implemented through tile drainage, where field subsurface drainage was installed in poorly drained soils (USDI, 1988). In the Southeast and Gulf coast states, reclamation was implemented mainly by moderating annual flood events by diking off direct linkages to rivers and streams, damming rivers to regulate peak flows and by mechanically pumping surface and subsurface water levels down to maintain conditions suitable for agriculture. During this period, government agricultural subsidies that reduced the financial risk of crop loss in flood prone areas accelerated wetland conversion (USDI, 1994). This period of wetlands conversion is often termed the "Era of Wetland Exploitation" (Heimlich *et al.*, 1998) when wetlands were thought of as wastelands and their usefulness to a growing and industrialising country was conversion to lands that could profit landowners.

Between the 1780's and 1954 wetlands were being converted to other land uses at an annual rate of 330,000 hectares or more (Pavelis, 1987; Dahl, 1990). Since the mid 1950's rates of conversion have decreased significantly (Table 1). However, these numbers only account for total conversion by drainage or infill, to this day these numbers do not quantify changes in quality often resulting from development of surrounding land use that can directly and indirectly impact wetland functions and values. Even with this rate of wetland loss in decline since the mid 1950's, an estimated 114 to 124 million acres of wetlands have been lost throughout the lower contiguous United States. This loss was not proportionate across the country. Twenty two states lost 50% or more of their wetlands, and of those, seven states lost 80% or more of their wetland area. Two states, California, and Ohio lost more than 90% of their predevelopment wetland area. Those states with the largest acreage of wetlands, Louisiana and Florida, have lost less than 50% of their original wetlands, but the total acreage is significant, 3.79 million hectares in Florida and 2.99 million hectares in Louisiana.

Table 1. Estimated wetland acres lost between settlement and 1992 in the lower 48 contiguous states.

Period of record	Average annual wetland area lost (hectares)	Source
Settlement - 1954	329,420-358,960	Dahl, 1990
1954 - 1974	185,350	Frayer *et al.*, 1983
1974 - 1982	117,360	Dahl and Johnson, 1991; Heimlich and Melanson, 1995
1982 - 1992	31,970	USDI-USGS 1996; Heimlich and Melanson, 1995

Wetland functions

In addition to area loss of wetlands many ecological and physical functions, which wetlands provide within a landscape were also lost. Wetland functions can be defined as any process, biotic or abiotic, that occur in a wetland that can be linked to a larger action (Novitski *et al.*, 1996; Williams, 1996). Biogeochemically, wetlands help maintain and improve water quality through uptake of nutrients by plant life, sorption to sediments and deposition of detritus and chemical precipitation. Chemistry of wetlands is a complex interaction between aerobic and anaerobic biogeochemical processes often resulting in the transformation and removal of toxic and/or undesirable compounds from the water column by microbial, vegetation, and/or sediment retention. Vegetation and the low topographic relief in wetlands typically slows water, reducing energy, thereby causing sediment deposition, which reduces downstream siltation of rivers, lakes, streams and estuaries. Some wetlands, such as riparian wetlands provide wildlife corridors, as they often occur along linear landscape features such as streams and rivers. These corridors provide a connection between otherwise fragmented habitat in the landscape. Ecologically wetlands also increase landscape floral and faunal diversity providing a transitional zone between upland areas such as agricultural grassland and tillage crops and aquatic fresh and saline water environments. Biologically wetlands are the most productive ecosystems in temperate regions rivalling rain forests in the tropics (Mitsch and Gosselink, 1993). Wetlands provide habitat for fish and wildlife, some species spend their entire lives in wetlands, while others use them intermittently for feeding, resting and/or breeding phases of their lives (Kroodsma, 1979). Aquatic species often depend on wetlands as a nursery ground for juvenile fish where protection and food supply are abundant. An estimated 95% of commercially harvested marine fish and shellfish are dependant on wetlands at some point in their lifecycle (Mitsch and Gosselink, 1993). Amphibians and reptiles are especially dependant on wetlands for habitat as many species depend on both terrestrial and aquatic phases in their lifecycle. In North America, over one-third of all bird species rely on wetlands for migratory resting places, breeding or feeding ground (Figure 1).

This number of dependant bird species is disproportionate to the 5% of the landscape that remains as wetlands. In addition, many fur bearing animals, such as muskrat, beaver, otter, mink and raccoon prefer wetlands as habitat. Hydrologically wetlands provide several functions. As a result of their landscape position, wetlands often increase the detention time of surface water in the landscape increasing the likelihood of groundwater recharge and localised evapotranspiration. Wetland water storage abates peak stormwater flows (Figure 2), reducing the total discharge volume to rivers and streams and thus, reducing the velocity of water that can cause in stream bank erosion and river bed scour (Carter, 1996).

Figure 1. Wetlands providing critical habitat for White Ibis and Roseate Spoonbill, two species of special concern in Florida. (Photo courtesy of South Florida Water Management District).

Nutrient management in agricultural watersheds: A wetlands solution

Figure 2. Floodplain water storage potential indicated by the height of lichen lines on tree trunks. Preserved and restored floodplains reduce peak flow during storms and extend baseflow runoff conditions.

All of these functions are provided "free" by intact wetlands ecosystems; however, that does not necessarily mean that they are valued by landowners and/or society.

Period of wetland enlightenment

The terms function and value may or may not be synonymous. As defined earlier, wetland functions are those processes or actions that occur in a wetland such as habitat for wildlife, biogeochemical transformation of nitrate to nitrogen gas, and sequestration of carbon from the atmosphere to name a few. Wetland values on the other hand are humanly defined and dependant on the perspective of an individual or society (Barbier *et al.*, 1977). Education, research and public policy promote the linkages and influence perception and therefore perceived value by individuals. But other factors such as religion, economic condition and philosophical viewpoint may be of equal or greater influence on how wetland functions are valued. Therefore, functions of a wetland are natural processes that exist regardless of their perceived value to an individual or society, whereas wetland values are typically utilitarian and associated with goods and services that wetlands provide directly or indirectly within the landscape (Table 2). As we have increased our understanding of the function of wetlands, society's value of these ecosystems has changed considerably (Carey *et al.*, 1990; Dahl and Allord, 1996).

With this change in public opinion, government policy also changed, and instead of programs promoting the conversion of wetlands to "productive lands" for agriculture and urban development, many programs directly or indirectly have more recently focused on the conservation and preservation of wetlands, as well as the restoration of wetlands throughout the United States. One of the first organisations to see the multifunctional benefits of wetland ecosystem conservation were duck hunters. They realised the critical habitat wetlands

Table 2. Examples contrasting differences between wetland functions and societal values often attributed to wetlands.

Wetland Functions	Wetland Values
Water storage in floodplains and depressional wetlands	– Peak flood attenuation
	– Extend groundwater recharge potential
	– Maintain baseflow in streams and rivers
Aquatic species habitat	– Commercial and recreational fisheries
	– Refugia for rare and threatened species
Terrestrial and waterfowl species habitat	– Wildlife viewing and hunting
	– Refugia for rare and threatened species
Sedimentation, pathogen retention, denitrification, and phosphorus burial in soil	– Maintain and improve water quality
Carbon sequestration and burial	– Partial regulation of global carbon budget
Vegetative stabilisation of soil by roots	– Reduce shoreline erosion and impacts of storm events

provided for recreational and subsistence hunting of waterfowl. In 1934 the Migratory Bird Hunting Stamp Act was past, funds of which were used to acquire wetlands principally in major flyways to preserve waterfowl habitat. Although the principal objective of this act was to conserve habitat for a consumable resource, significant benefits to water quality and non game species habitat were also conserved and protected in the process. In 1970 the Water Bank Program provided annual per acre payments to farmers with eligible wetlands not to put these wetlands into agricultural production. This same programme is now the most successful wetland preservation/restoration program targeting wetlands. It is presently titled the "Wetland Reserve Program," (WRP) and is administered by the Natural Resource Conservation Service. Arguably the most influential programme to protect wetlands came as part of the 1972 Clean Water Act. This established ridged delineation and permitting requirements that promoted, "avoidance, minimisation and mitigation" for dredge or fill activates in wetlands. It also protected wetlands against water quality related impacts and is presently facilitating the establishment of numeric nutrient criteria for surface waters that will set nutrient standards that are thought protective of the ecological integrity of wetland ecosystems. In 1977 an executive order under President Carter directed all federal agencies to minimise the destruction, loss, or degradation of wetlands associated with any federal activity. As a partial result of that order, the Swampbuster Provision in 1985 eliminated indirect farm program benefits and tax deductions to farmers who drained protected wetlands. In 1991 and 1993 presidents Bush and Clinton both used the iconic and often controversial phrase "No Net Loss" to promote wetland mitigation and attempt replacement of wetlands for any wetland losses incurred (White House 1991, 1993). Although this term promotes the concept of equal replacement of wetland loss, there is still much debate as to the currency of "equal" replacement and degree of success (Steinhart, 1987; NRC 1992; Kusler and Kentula, 1990; Kentula, 1996; Hunt, 1996). For instance, should replacement occur simply in acreage loss? Is it important to have the same kind of wetland that was lost? What about the location of a wetland in a watershed does it matter? How does one account for inequality between the functions lost from a wetland that developed along a geological time scale with one created on a time scale of months to years?

Not all wetland protection and creation are dictated by policy. Natural processes at work in wetlands can also provide a means for agricultural interest to address and comply with increasingly stricter water quality regulations imposed by federal and state regulatory agencies.

A new paradigm for wetlands

Historically agricultural interest viewed wetlands as unproductive lands that were beneficial only if drained and converted to tillage or grazing lands. The economic bottom line was only served if these lands could provide some form of income. To the rest of society, wetlands were valued to preserve water quality, provide habitat for fish and wildlife, prevent erosion, reduce flood damage and provide aesthetically pleasing open space and recreational areas. These societal viewpoints resulted in many of the policies that individual landowners must comply to. But how does one address the differences between private interests and public values? Many groups benefit from unimpacted wetland functions: anglers, hunters, boaters, downstream property owners, public water supply and flood control authorities among others. However, private owners usually can not benefit from these values economically unless the wetlands are converted or maintained as other uses, such as agriculture or urban development (Alvayay and Bean, 1990; Kohn, 1994; Heimlich *et al.*, 1998). The biogeochemical processes that improve water quality in wetlands may change the value of wetlands for agricultural interest. In response to new and upcoming regulations and more stringent water quality discharge requirements, innovative technologies and implementation of Best Management Practices (BMPs) may lead to a new wetlands paradigm for landowners.

The same biogeochemical functions present in natural wetlands can also assist land owners in meeting stricter water quality criteria set forth at the state and federal levels. Integrating wetlands in agricultural landscapes can provide societal values and reduce agricultural impacts to water quality. In many respects this has changed the agricultural value of wetlands from wastelands to a BMP allowing these private interests to comply with stricter water quality expectations (Figure 3).

In some instances, implementation of recommended BMPs has lead to a "presumption of compliance" with water quality standards, while effectiveness at the watershed scale is evaluated (LOPP, 2004). Evidence suggests that integration of wetlands can provide a landowner with significant benefits (Kadlec and Knight, 1996; Knight *et al.*, 1993). However, the bottom line still remains where the landowner can not recoup any economic benefits from a wetland in this capacity, and although this means of water quality improvement may be less costly than a more intensive treatment system, the benefits associated with water quality improvement are often perceived by the landowner as a societal benefit that should not burden the landowner. This argument is only partially true, as water quality impacts would not exist if agriculture was not first causing landscape modification, so indeed some of the "burden" to improve water quality must be born by the landowner. However, society also requires the services of agriculture to provide food and therefore certain realities of agricultural practices, including intensification and fertilisation of the landscape must also be assumed by society. Therefore, food production and clean water are issues that both agricultural landowner and society must address and share in the burden/cost.

Figure 3. Tree, shrub and native grass wetland buffer in Iowa reduces upland runoff, improves stream water quality and provides wildlife habitat. (Photo courtesy of USDA NRCS).

In a true market economy, any costs associated with production would be transferred from the producer to the consumer of the product. Thus, if additional cost associated with integrating a wetland into an agricultural operation for purposes of meeting new water quality standards, those costs would be passed onto the consumer. However, in a global economy, cost of environmental compliance is not always spread equally and in areas with stricter environmental regulations the cost of a product will often be higher than in areas that do not have mechanisms for environmental protection. Beyond the local scale of exchange, processors will simply change suppliers to reduce their cost and shift the environmental impacts elsewhere to less protected areas. To minimise this relocation of environmental impacts, localised markets within watersheds that take into account the cost of environmental protection and enable cost transfer directly to the consumer would need creation. However, this level of intervention, although it may be a more sustainable solution, requires a re-evaluation of an increasingly "global market economy" at least with regard to agricultural products. An alternative to this economic intervention is to subsidise the cost of environmental protection through "cost sharing" government and non government programmes that spread the monetary burden between private and public interests (USACE, 1994).

Cost share programmes and means to incentivize wetland protection and their integration into agricultural management practices are presently the most successful strategies to enhance and protect wetlands on agricultural lands. The most successful of these programmes is the WRP as noted earlier. Goals of the WRP are to achieve the greatest wetland function and values along with optimum wildlife habitat on every acre enrolled. There are multiple durations of commitment by an agricultural land owner; 10 year, 30 year and permanent easements. Currently, there are over 400,000 hectares enrolled in the program with a reauthorisation in 2002 of the Act to include up to 918,650 hectares. With the exception of non government programmes, funding of these cost share programmes is from taxation at either a federal, state or local level. They provide landowners with an economic incentive to maintain, restore or create wetland areas to improve water quality, store water, and provide habitat, while benefiting society by maintaining the broader benefits and values that we have come to associate with wetlands. The economic resource base to support theses programmes, however, is finite and tax supported subsidies to maintain wetlands on agricultural landscapes would at some point be better served by more direct linkages between production and consumer costs.

Conclusion

In summary, significant changes in society's values of wetlands has occurred in the past fifty years and as a result increased environmental awareness and protective policy has been initiated and implemented. Hydrological, ecological, biological, physical and chemical functions and their values are now realised by the public and private sectors alike. The majority of wetlands in the US lie in private land ownership and therefore wetland ecosystem protection requires a shared burden between landowner interest and society benefits by preserving their function. "Cost share" programmes provide a near-term mechanism to incentivize this integration without undue monetary burden placed on the agricultural landowner. Wetlands are not the panacea to water quality issues in developing watersheds, but can provide an ecologically sensitive means to improve water quality concerns at the watershed scale and provide multifunctional benefits for both private and public interests.

References

Alvayay, J., and J.S. Baen, 1994. Wetland Regulation in the Real World. Beveridge and Diamond, Washington, D.C., 27 pp.

Barbier, E.B., M. Acreman, and D. Knowler, 1997. Economic Valuation of Wetlands: A Guide for Policy Makers and Planners. Ramsar Convention Bureau, Department of Environmental Economics and Environmental Management, University of York, Institute of Hydrology, IUCN-the World Conservation Union, Cambridge, England, 135 pp.

Carey, M.B., R.E Heimlich, and R.J. Brazee, 1990. A Permanent Wetlands Reserve: Analysis of a New Approach to Wetland Protection. AIB-610, US Department of Agriculture, Economic Research Services.

Carter, V. 1996, Wetland Hydrology, Water Quality, and Associated Functions. In: National Water Summary on Wetland Resources, edited by J. D. Fretwell, J.W. Williams, and P.J Redman, USGS Water Supply Paper 2425. USDI, US Geological Survey, Washington, DC, pp. 35-48.

Dahl, T.E. 1990, Wetlands losses in the United States, 1780's to 1980's. US Department of the Interior, U. S. Fish and Wildlife Service. Washington, DC, 21 pp.

Dahl, T.E. and C.E. Johnson, 1991. Status and Trends of Wetlands in the Conterminous United States Mid-1970's to Mid-1980's. US Department of the Interior, Fish and Wildlife Services, Washington, DC, 28 pp.

Dahl, T.E., and G.J. Allord, 1996. History of Wetlands in the Conterminous United States. In: National Water Summary on Wetland Resources, edited by J. D. Fretwell, J.W. Williams, and P.J Redman, USGS Water Supply Paper 2425. USDI, US Geological Survey. Washington, DC, pp.19-26.

Frayer, W.E., T.J. Monahan, D.C. Bowden, and F.A. Graybill, 1983. Status and Trends of Wetlands and Deepwater Habitats in the Conterminous United States, 1950's to 1970's. Fort Collins, CO, Colorado State University, 31 pp.

Heimlich, R.E. and J. Melanson, 1995. Wetlands Lost, Wetlands Gained, National Wetlands Newsletter **17** (3)1-25.

Heimlich R.E., K.D. Wiebe, R. Claassen, D. Gadsby, and R.M. House, 1998. Wetlands and Agriculture: Private Interests and Public Benefits. Agricultural Economic report No.765, US Department of Agriculture, Economic Research Service. 94 pp.

Holtman, C., S.J. Taff, A. Meyer, and J.A. Leitch, 1996. An Inquiry into the Relationship of Wetland Regulations and Property Values in Minnesota. Staff Paper P96-16. Department of Applied Economics, University of Minnesota, St. Paul, Dec., 32 pp.

Hunt, R.H. 1996. Do Created Wetlands Replace the Wetlands that are destroyed? FS-246-96, US Department of the Interior, US Geological Survey, 4 pp.

Kadlec, R.H. and R.L. Knight, 1996. Treatment Wetlands. Lewis Publishers, Boca Raton, FL, 893 pp.

Kentula, M.E. 1996. Wetland Restoration and Creation. In: National Water Summary on Wetland Resources, edited by J.D. Fretwell, J.W. Williams, and P.J Redman, USGS Water Supply Paper 2425. USDI, US Geological Survey. Washington, DC, pp. 87-92.

Knight, R.L., R.W. Ruble, R.H. Kadlec, and S.C. Reed, 1993. Wetlands for Wastewater Treatment Performance Database. In: Constructed Wetlands for Water Quality Improvement, edited by G.A. Moshiri, Lewis Publishers, Boca Raton, FL. 632 pp.

Kohn, R.E. 1994. Alternative Property Rights to Wetland Externalities. Ecological Economics **10** 61-68.

Kroodsma, D.E., 1979. Habitat Values for Nongame Wetland Birds. In: Wetland Functions and Values-The state of Our Understanding, edited by P.E. Greeson, Clark, J.R., and Clark, J.E., Minneapolis, MN: American Water Resources Association, pp. 320-343.

Kusler, J.Q., and M.E. Kentula, 1990. Wetland Creation and Restoration: The Status of the Science. Washington, DC: Island Press, 591 pp.

Lake Okeechobee Protection Program (LOPP), 2004. Lake Okeechobee Protection Program, Annual Report to the Florida Legislator, South Florida Water Management District, Florida Department of Environmental Protection, Florida Department of Agriculture and Consumer Affairs, January, 63 pp

Mitsch, W.J. and J.G.Gosselink, 1993. Wetlands. Van Nostrand Reinhold, New York, 722 pp.

Novitski, R.P., R.D. Smith, and J.D.Fretwell, 1996. Wetland Functions, Values, and Assessment, In: National Water Summary on Wetland Resources, edited by J.D. Fretwell, J.W. Williams, and P.J Redman, USGS Water Supply Paper 2425. USDI, US Geological Survey. Washington, DC, pp. 79-86.

National Research Council (NRC), 1992. Restoration of Aquatic Ecosystems: Science, Technology, and Public Policy. National Academy Press, Washington, DC. 576 pp.

Pavelis, G.Q., 1987. Economic Survey of Farm Drainage: In: Farm Drainage in the United States, edited by G.A. Pavelis, US Department of Agriculture, Economic Research Service, Washington, DC. 170 pp.

Shaw, S.P., and C.G. Fredine, 1971. Wetlands of the United States: their Extent and their Value to Waterfowl and other Wildlife. Circular 39, US Department of the Interior, Fish and Wildlife Service, Washington, DC, 67 pp.

Steinhart, P., 1987. Mitigation Isn't. Audubon **89** 8-11. May.

United States Army Corps of Engineers (USACE), 1994. National Wetland Mitigation Banking Study: an Examination of Wetland Programs: Opportunities for Compensatory Mitigation. Institute for Water Resources Report 94-WMB-5, 63 pp.

Unites States Department of the Interior (USDI), 1988. The Impact of Federal Programs on Wetlands, Volume 1. The Lower Mississippi Alluvial Plain and the Prairie Pothole Region. A Report to Congress by the Secretary of the Interior. Oct., 114 pp.

United States Department of the Interior (USDI), 1994. The impact of Federal Programs on Wetlands, Volume 2. The Everglades, Coastal Louisiana, Galveston Bay, Puerto Rico, California's central Valley, Western Riparian Areas, Southeastern and Western Alaska, The Delmarva Peninsula, North Carolina, Northeastern New Jersey, Michigan, and Nebraska. A Report to Congress by the Secretary of the Interior. Mar., 333 pp.

United States Department of the Interior, US Geological Survey, 1996. National Water Summary on wetland Resources, edited by J.D. Fretwell, J.S. Williams, and P.J. Redman, USGS Water Supply Paper 2425.

White House Office on Environmental Policy. 1993. Protecting America's Wetlands: A Fair, Flexible, and Effective Approach. Aug. 24, 26 pp.

White House Office of the Press Secretary, 1991. Fact Sheet: Protecting America's Wetlands. Press release dated Aug. 9, 5pp.

Williams, T. 1996. What Good is a Wetland? Audubon **98** (6):42-53. November-December.

Wetland restoration within agricultural watersheds: balancing water quality protection with habitat conservation

A. Niedermeier[1], J.S. Robinson[1] and D. Reid[2]
[1]Department of Soil Science, School of Human and Environmental Sciences, The Universiy of Reading, Berks, RG6 6DW, UK
[2]The Somerset Wildlife Trust, Bridgwater Fyne Court, Broomfield, Bridgwater, Somerset, TA5 2EQ, UK

Abstract

Peat wetlands that have been restored from agricultural land have the potential to act as long term sources of phosphorus (P) and, therefore have to potenital to accelerate freshwater eutrophication. During a two-year study the water table in a eutrophic fen peat that was managed by pump drainage fluctuated annually between +20 cm and -60 cm relative to ground level. This precise management was facilitated by the high hydraulic conductivity (K) of the humified peat (1.1×10^{-5} m s^{-1}) below around 60 cm depth. However, during one week of intermittent pumping, as much as 50 g ha^{-1} dissolved P entered the pumped ditch. Summer rainfall events and autumn reflooding also triggered P losses. The P losses were attributed to the low P sorption capacity (217 mg kg^{-1}) of the saturated peat below 60 cm, combined with its high K and the reductive dissolution of Fe bound P.

Keywords: Phosphorus, peat, wetland, water.

Introduction

The restoration of wetlands as bird habitats involves the maintenance of a suitable water regime, appropriate vegetation types and invertebrate populations. In response to the UK government's agri-environment programme, a growing number of restoration efforts for birds are on peat-based, lowland wet grasslands that had previously undergone agricultural improvement by drainage and fertiliser applications. In most cases, the water regime on these sites is maintained by careful, localised ditch water management using pumps and sluices. However, the alternate wetland/dryland cycles that define the water regime of many restored peatlands can accelerate nutrient cycling within the peat substrate. As a consequence, particularly on former arable sites, considerable quantities of the nutrients (mainly nitrogen and P) may be transformed from chemically and biologically stable forms to labile forms, thereby posing a threat to ditch water quality through the process of eutrophication. Moreover, the accelerated process of eutrophication delays the progress of restoration projects owing to the requirement of wetland birds for a wide range of vegetation types.

The study reported here addresses the dynamics and losses of soil P in a recently re-wetted, eutrophic fen peat. The specific objectives were to evaluate the relationships between soil hydrology, hydrochemistry and P dynamics in a small wetland in which the water levels are managed for birds.

Material and methods

Site description

The Catcott Lows Wetland Reserve covers approximately 50 ha of the Somerset Levels and Moors Environmentally Sensitive Area (ESA) in Southwest England and is centred on grid reference ST 404416 (Figure 1). The Reserve consists of six fields of wet grassland with negligible hydraulic gradient that lie mostly below 1.8 m O.D. Shortly after acquisition in 1990, the Somerset Wildlife Trust (SWT), Bridgwater, blocked all underground drainage pipes and reseeded the arable fields with grass.

The research reported here focuses on Lot 4 (Figures 1 and 2). Lot 4 covers an area of 7.8 ha and lies in a zone of Altcar 1 Series Fen peat, which is an organic soil of predominantly eutrophic peat, overlying marine clay (Findlay *et al.*, 1984). The field water regime is managed by the SWT through control of the water levels in the small drainage ditch on the west border of the Lot (West Ditch) by the operation of a diesel pump (Figure 2). A supplementary pump and sluice are also installed where the Reserve's main east-west ditch links with Lady's Drove Rhyne (Figure 1). The much larger East Ditch (or *Black Ditch*) that runs along the east side of the field is hydrologically isolated from the Reserve. A summary of the water levels that are targeted by the SWT for Lot 4 is provided in Table 1. This water regime aims to maximise the site's value as a wintering ground for migrant wildfowl and waders (as a grazing marsh), and its value as a breeding ground for waders, whilst improving the vascular plant diversity. During the late spring and/or summer, depending on rainfall, West Ditch may need to be pumped intermittently to remove excess water from the field. This ensures that the ground is dry enough for breeding waders to nest in June and also that the land surface is sufficiently dry in July/August to harvest the vegetation.

Soil physicochemical properties

Soil samples were taken from Lot 4 in October 2000, when the water table was approximately 15 cm below the land surface. Soil cores were taken to 150-cm depth at 20 locations along a 'W' shaped path using an extendable cheese-type corer. The cores were divided into the following layers: 0-26 cm, 26-50 cm, 50-100 cm and 140+ cm on the basis of the variation

Table 1. Water level management plan for Lot 4.

Time of year	Field water level (O.D.) (metres)
October to end of January	1.4 to 1.6
During February	1.7
mid March to early May	1.65
end May to mid June	1.45
mid/late July	1.1
August to September	1.3

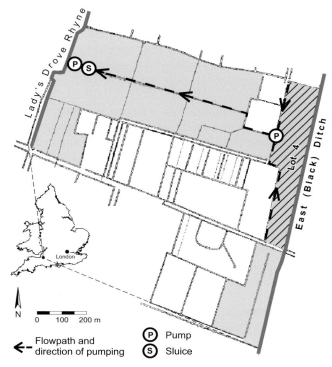

Figure 1. Site location and water management.

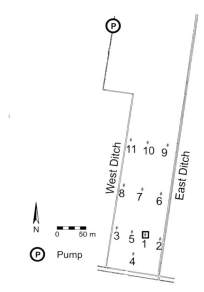

Figure 2. Arrangement of dipwell stations and location of Station 1 in Lot 4.

in colour. Each layer was further characterised in the field for *von Post* class and any other visible characteristics. At each of the twenty locations, the hydraulic conductivity (*K*) of the peat was measured in each peat/clay layer, *in situ*, using a rising auger hole method. Duplicate samples for the determination of bulk density were taken horizontally from the face of a soil pit using a 5-cm i.d. metal cyclinder (Rowell, 1984).

For determination of soil chemical properties, representative samples were taken from the 0-30, 30-60 and 60-90 cm layer; these depths were chosen to align with the depths, 30, 60 and 90 cm, at which the field was instrumented for hydrological and hydrochemical dynamics, as described in the section that follows. The soil samples were bulked in to the different layers, returned to the laboratory, air dried and passed through a 4-mm sieve to remove any discernible plant roots. Soil organic matter was estimated as loss-on-ignition at $550^{\circ}C$ (Nelson and Sommers, 1982), and pH was measured in 0.01 M $CaCl_2$ at a soil:solution ratio of 1:2.5 (Rowell, 1994). Olsen P was extracted from 1 g soil by 20 ml 0.5 M $NaHCO_3$ (pH 8.5) for 30 mins (Olsen *et al.*, 1954). Total P, calcium (Ca) and iron (Fe) contents were determined after digestion of oven-dried samples in a nitric acid-perchloric acid mixture. All extracts were filtered through a Whatman No. 42 filter paper prior to analysis for P, Ca and Fe by inductively coupled plasma atomic emission spectroscopy (ICP-AES) analysis.

Phosphorus sorption characteristics were determined on the 0-30, 30-60 and 60-90 cm depth increments of the soil, using a method adapted from Nair *et al.* (1984). Langmuir P sorption isotherms were fitted for the calculation of P sorption parameters: soil P sorption maximum and soil P binding affinity (Syers *et al.*, 1973).

Field instrumentation

In November 2000, dipwells were installed in the field to monitor water table fluctuations. The dipwells were installed at 11 stations on a rectangular grid pattern in the southern half of Lot 4, covering an area of approximately 3.5 ha (Figures 2 and 3). At each station, the elevation was measured using a theodolite (Geodimeter 410). One of the stations (Station 1, 160 cm O.D.) was equipped with further instrumentation for monitoring soil water suction and redox potential (Figure 3). Field tensiometers were installed at the 30, 60 and 90-cm soil depths. The suction probes were calibrated to 0 kPa suction when the soil was saturated to the height of the porous ceramic cup. Redox probes were installed at depths from 30 to 100 cm in increments of 10 cm. An Ag/AgCl reference electrode in combination with platinum electrodes was used to measure the redox potential. The soil water suction and redox values were recorded continuously from January 2001 to December 2002 at one hour intervals on a data logger (Figure 3).

Drainage ditch management and measurements

In order to meet the management plan, the SWT intermittently pumped West Ditch down to 120 cm O.D. (approximately 40 cm below the land elevation at Station 1) during June to August 2001 and August to September 2002. On June 7th, 2001 the pump was operated for an initial eight hour event and for subsequent eight hour events on two occasions when rainfall flooded the surface of significant portions of the field (4th July and 10th August). In 2002, the initial

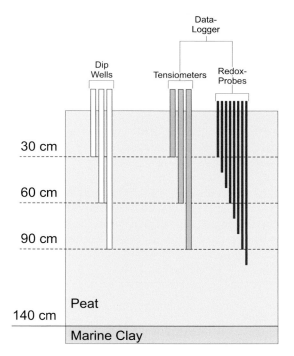

Figure 3. Instrumentation at Station 1.

eight hour pump event was on the 14th August, followed by a short series of frequent pump events for experimental purposes, each lasting three to six hours, on the 15th, 16th and 21st August. The final pump event of eight hours was on the 9th September in response to intense rainfall.

At monthly intervals, the ditch water level was recorded relative to the land elevation at Station 1. At the same periods, the water was sampled at four locations along its length, filtered to 0.45 µm and analysed for dissolved P (DP) using the colorimetric method of Murphy and Riley (1962). In 2002, more frequent measurements of water level and DP were made around the period (August - September) when West Ditch was pumped.

Meteorological measurements

Daily precipitation data were obtained from Goldcorner Meteorological Station (Grid reference ST 4309 3677). Potential evapotranspiration (ET_{pot}) was estimated on a monthly basis using the Penman model. Calculations were executed using the software DailyET v.2.1. from Cranfield University and utilised the following local measurements: temperature, humidity, sunshine hours and wind speed. As records for the year 2002 are incomplete, those for 2001 were used for both years.

Results and discussion

Soil physicochemical properties

The *von Post* classification indicates that the peat in the 60-90 cm layer is very humified, probably owing to the former deep drainage of the subsoil for arable agriculture (Table 2). The grainy, macroporous structure is consistent with the very low bulk density and high hydraulic conductivity of the 60-90 cm layer. The underlying, dense marine clay acts as a lower hydraulic boundary layer, having a hydraulic conductivity up to almost three orders of magnitude lower than the overlying peat.

Typical of eutrophic fen peats, the soil displayed circum-neutral pH values as well as high contents of soil organic matter (LOI), and total Ca and P (Table 3). The high Ca status is attributed to the Ca-rich water draining into Lady's Drove rhyne from the Lias Limestone Polden Hills to the south of the Reserve. Total and available (Olsen) P decreased with increasing depth. The relatively high P status of the upper soil layer reflects the agricultural fertiliser history of the former arable field.

Table 2. Selected physical properties of the peat.

Depth (cm)	Colour	Description	von Post class	Bulk density $(g\ cm^{-3})$	Hydraulic conductivity $(m\ s^{-1})$	Other visible characteristics
0-26	Black	Semi-humified, clayey peat	H6	0.34	5.9×10^{-7}	Platy clay structure; dense, live root mat.
26-50	Brown/Black	Semi-humified peat	H6	0.13	8.3×10^{-7}	Sharp decrease in root activity below 40 cm; sedge remains.
50-100	Dark brown	Grainy, macroprous peat	H9-10	0.083	1.1×10^{-5}	Very soft, degraded structure.
140+	Blue-grey	Marine clay	N.A.	1.60	2.2×10^{-8}	

Table 3. Selected chemical properties of the eutrophic fen peat.

Property	Soil depth (cm)		
	0 - 30	30 - 60	60 - 90
Loss on ignition (LOI) [g kg^{-1}]	589	903	880
pH	7.2	7.1	7.0
Olsen P [mg kg^{-1}]	30	8.8	9.2
Total P [mg kg^{-1}]	1100	254	302
Total Fe [g kg^{-1}]	3.5	2.1	2.4
Total Ca [g kg^{-1}]	20	16	15

The higher P sorption maximum and P binding affinity in the 0-30 cm layer compared with the lower soil depths can be attributed to the higher content of clay in the upper layer (Table 4). The Fe and Al oxide component, especially the amorphous forms that are characteristic of hydromorphic soil layers, have a particularly high sorption capacity for both inorganic and organically bound P. However, processes associated with the humification of the lower peat layers (> 60 cm) would have been conducive to the formation of humic acids that have a strong affinity for cations. As a consequence, Fe, Al and Ca ions are effectively blocked by the humic acids from fixing soil P; this may increase the vulnerability of P losses to drainage waters.

Table 4. Langmuir-derived soil P sorption maximum and P binding affinity for the 0-30, 30-60 and 60-90 cm soil depths.

Parameter	Soil depth (cm)		
	0 - 30	30 - 60	60 - 90
P sorption maximum [mg kg^{-1}]	2032	568	217
P binding affinity [l kg^{-1}]	0.0012	0.00015	0.00015

Hydrological fluctuations

A detailed profile of the fluctuations in field water level at Station 1 during the two-year study was provided by the automatically logged soil water tension (matric potential) at the 30, 60 and 90 cm soil depths (Figure 4). The scientific justification for deriving water table height from the tensiometry lies in the fact that the difference between the matric potential readings at the three depths remained constant at 3 kPa (30 cm tension) throughout the study. Transforming the matric potential to total water potential allows the tension readings at any of the three depths to be used to estimate the position of the water table at Station 1; for example, if the matric potential at 30 cm depth is 0 kPa, it is assumed that the soil water level (water table) is 30 cm below the surface. In fact, there was a good correlation (r = 0.97, $p < 0.001$) between monthly values for the derived soil water potential and observed water table height at Station 1 (data not presented), adding support to the use of soil tensiometry as a temporally precise surrogate for field water table position. Moreover, transforming the matric potential to total water potential enabled a visual comparison of the ditch and soil water levels (water table height) in response to pumping and to meteorological patterns (Figures 5 and 6).

At Goldcorner Meteorological Station a total of 643 mm precipitation was recorded in 2001; 53% less precipitation than that (984 mm) recorded in 2002. The start of the 'drained' period for each year is defined by the first day of pump drainage. Hence, the drained periods for 2001 and 2002 started on 7th June and 14th August, respectively (Figures 5 and 6). The end of the drained period is marked by the first precipitation event, following cessation of the pump drainage period, that brings the field water table up to within about 10 cm of the land surface. During most of the two drained periods, the field water table was more than 30 cm below ground level. For the purposes of the current study, a precipitation event is defined as a period during which precipitation occurs either without interruption or with interruptions that,

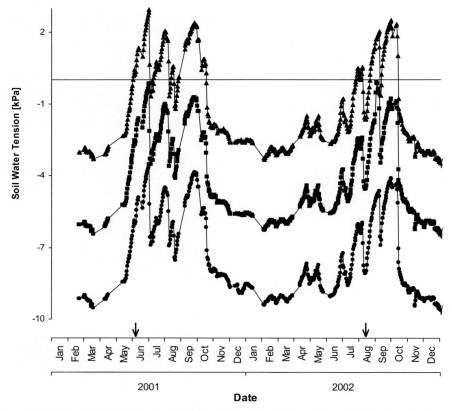

Figure 4. Variation in soil water tension at the 30 cm (■), 60 cm (●) and 90 cm (◆) depths at Station 1. The arrows indicate the periods of pump drainage of West Ditch.

individually, last no longer than five hours. In other words, precipitation that is interrupted by a period of longer than five hours constitutes two precipitation events; two periods of interruption, each greater than five hours, constitutes three precipitation events, and so on.

During the drained period for 2001 (from 7th June to around 20th October) 302 mm of rain were recorded, which is equivalent to 2.2 mm d^{-1}. This rainfall was spread evenly throughout most of the drained period, with the majority of rainfall events not exceeding 15mm d^{-1}. However, two events were extremely intense, at 45 and 33 mm d^{-1} on 4 July and 10 August, respectively (Figure 5). During the relatively short drained period for 2002, only 116 mm rainfall were recorded. However, similar to 2001, there were two outstandingly high rainfall events during the drained period of 2002 (Figure 6). The first intense event occurred on 9 September (total of 35 mm); the second event occurred on 14 October (total of 54 mm), marking the end of the drained period.

The close agreement between the rates of change in soil water level (measured at Station 1) and West Ditch water level (at least 56 m from Station 1) in response to the onset of pump

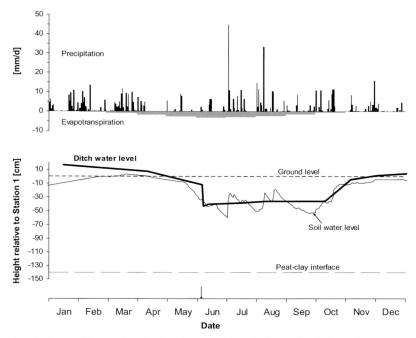

Figure 5. Meteorological conditions and fluctuations in West Ditch and soil water levels during 2001. The arrow indicates the onset of pump drainage.

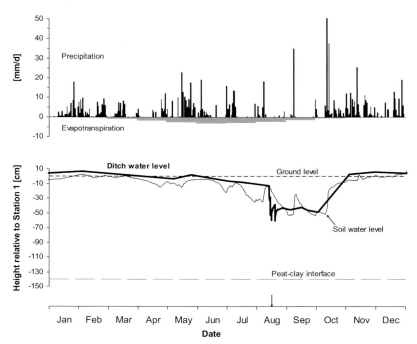

Figure 6. Meteorological conditions and fluctuations in West Ditch and soil water levels during 2002. The arrow indicates the onset of pump drainage.

Nutrient management in agricultural watersheds: A wetlands solution   *63*

drainage and of the subsequent autumn re-wetting in both 2001 and 2002 demonstrates that there is very good hydrological connectivity between the field and West Ditch (Figures 5 and 6). On the contrary, in a case study of the influence of water table drawdown in a valley fen peat field at West Sedgemoor (Somerset Levels), Heathwaite (1991) observed that drawdown was restricted to within 5 m of the ditch and attributed this to the low hydraulic conductivity (0.3 to 2.3 x 10^{-6} cm s^{-1}) of the peat. For upland peats, Hudson and Roberts (1982) could not detect a significant change in soil moisture beyond 1 m of the drainage ditch. Similarly, Stewart and Lance (1983, 1991) identify this zone as a few metres for blanket peats. The disagreement between the current and previous research may be attributed to the extremely high hydraulic conductivity (1.1 x 10^{-5} m s^{-1} - equivalent to approximately 1 m d^{-1}) of the humified peat below 50 cm depth (Table 2).

Redox dynamics

Redox potential measurements at all soil depths between 30 and 100 cm varied markedly over the course of the two years. Redox data data are shown only for the 30, 60 and 90 cm depths (Figure 7). The Eh fluctuations at the 40, 50, 70 and 80 cm depths followed very similar patterns and values to those shown at the 30, 60 and 90 cm depths. The most marked temporal variations in Eh were during 2001 (Figure 7). Prior to the drained period in 2001, Eh values at all depths remained below -100 mV, indicating anaerobic conditions throughout the whole soil profile. During the drained period of 2001, Eh values at all depths peaked at values between +100 and +230 mV, but displayed large temporal variation during this period. The detection limit of O_2 at a pH of 7 is reached at approximately +330 mV (Scheffer and Schachtschabel, 1998). Therefore, in spite of the relatively high Eh values during the drained period, it is likely that most of the soil profile remained anaerobic. At the end of the drained period, Eh at all depths by the end of 2001 had stabilised between +50 and -50 mV. In early February 2002, Eh reached approximately the same values as measured in the February of 2001 (-100 to -150 mV). During the drained period of 2002 (start 14th August), the fluctuations in Eh were less pronounced compared with those in 2001, and varied less among the different depths, ranging from -180 to -50mV (Figure 7). In spite of the intensive period of pump drainage that took place in August 2002, the whole soil profile remained anaerobic. The permanently reduced state of the soil profile greater than 30 cm depths throughout the two-year study is explained by the fact that the water table never fell below -60 cm (Figures 5 and 6). At a water table depth of 60 cm, it was estimated from laboratory determinations of air entry pressure that the capillary fringe would have extended through the mineral-peat matrix of the surface 0-50 cm to within about 30 cm of the soil surface (data not presented). In a fen peat, with similar profile characteristics and flood-drainage regime to those in Lot 4, Fiedler (2000) also found persistent anaerobic conditions below a depth of 10 cm.

In spite of the permanently reduced state of the soil profile from 30- to 100-cm depth, there was a strong, negative, linear relationship ($p < 0.001$) between daily values for redox potential at these depths and water table height; the time series for redox potential and soil water tension at 30-cm depth demonstrate this relationship (Figure 8). The general inverse trend of ground water level and Eh in periodically flooded soils is well documented among other workers (Faulkner et al., 1989) but examples and explanations of the relationship are limited to cases where the soil is periodically oxygenated. The trends presented here demonstrate that different

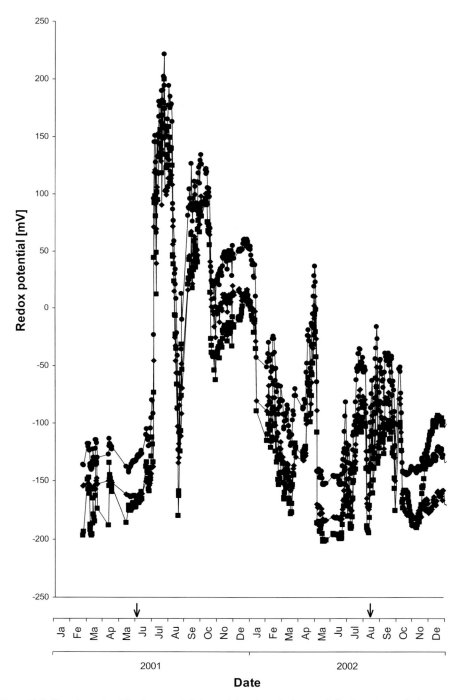

Figure 7. Daily redox potential values at 30 (■), 60 (●) and 90 (♦) cm soil depths at Lot 4 during 2001 and 2002. The arrows indicate the onset of pump drainage.

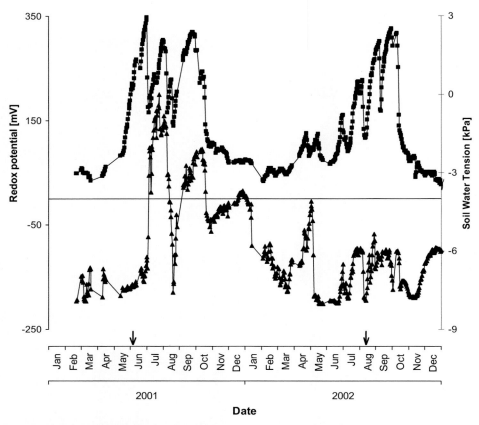

Figure 8. Relationship between redox potential at 30 cm soil depth (■) and soil water potential (▲). The arrows indicate the onset of pump drainage.

alternative electron acceptors were poising the Eh at various stages of the hydroperiod. For example, the reduced conditions during the flooded periods, the Eh values between -100 and -200 mV may have been poised by high concentrations of Fe(II) in equilibrium with a sequence of various amorphous Fe hydroxide precipitates (Khalid *et al.* 1977). It is possible that the reductive dissolution of the highly P retentive oxides and hydroxides of Fe contributed to increases in soil solution P (data not presented).

Phosphorus levels in West Ditch

Dissolved P concentrations in West Ditch varied markedly during the two-year period and appeared to fluctuate in response to water levels (Figure 9). In 2001, there were clear increases in ditch P concentration in response to the onset of pump drainage in June (up to 400 µg l^{-1}) and the re-wetting in October (up to 700 µg l^{-1}).

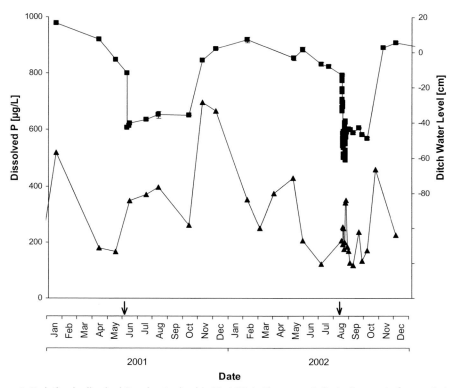

Figure 9. Variation in dissolved P and water level in West Ditch. The arrows indicate the onset of pump drainage.

The influence of pumping on the increasing P load in West Ditch was quantified for 2002, based on calculations of the estimated volumes of water pumped and the measured concentrations of P in the ditch. Approximately 1600 m^3 of water and 400 mg of dissolved P were pumped out of West Ditch during the 7 day study period. It is likely that the high hydraulic gradient, imposed by pumping, was responsible for the transfer of much of the P from the field to the ditch by mass flow. The process of diffusion would have played a relatively very minor role in the transfer of P owing to the deep peat layers having both a high porosity and saturated hydraulic conductivity (Table 2). Furthermore, the very weak P retention properties of the peat at these lower depths would have offered little resistance to the mass flow of dissolved P to the drainage ditch (Table 4).

Further evaluation of the intensively monitored period between mid-August (start of drained period) and November 2002 shows the possible influence of summer rainfall events on dissolved P concentrations in West Ditch as well as that of the August pumping (Figure 10).

The first peak in P concentration was in response to the onset of pumping; the second and third rapidly (within 4 to 5 days) followed the two intense summer rainfall events that occurred on the 9th September and 14th October. It is argued that the latter two P peaks were attributed to the mass flow of P from the field during the drainage events that immediately followed the

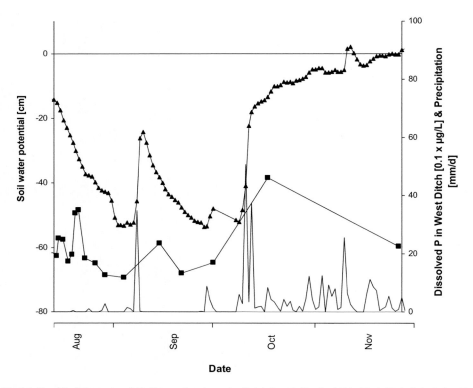

Figure 10. Relationship between precipitation, soil water potential (■) and dissolved P in West Ditch (▲) during the period from the onset of pump drainage in August 2002 to reflooding in November.

sharp rises in water table height that occurred in resonse to the rainfall events (Figure 10). Owing to the low air-filled soil porosity, 30-50 mm d^{-1} rainfall events rapidly raised the water table by up to 30 cm, to within 20 cm of the land surface. In other words, the two rainfall events rapidly created a positive hydraulic gradient between the field and the West Ditch, as demonstrated by the rises in soil water potential above the West Ditch water level (Figure 11). Similarly, positive hydraulic gradients were observed during the extreme rainfall events on 4th July and 10th August, 2001 (Figure 12). Similar to the effect of pump drainage, the rainfall-driven discharges of soil water probably carried high concentrations of weakly buffered P to the West Ditch from the strongly humified, deep peat layers.

Conclusions

During the two year study, the field water table of the restored wetland was managed successfully to fluctuate annually between +20 cm and -60 cm relative to ground level, thereby providing a suitable wintering ground for migrant wildfowl and as a breeding ground for waders. The water regime was strongly influenced by pump drainage of the adjacent ditch water level and was facilitated by the high hydraulic conductivity of the humified peat (1.1 x 10^{-5} m s^{-1}) below around 60-cm depth. Also with increasing peat depth there was a marked decrease

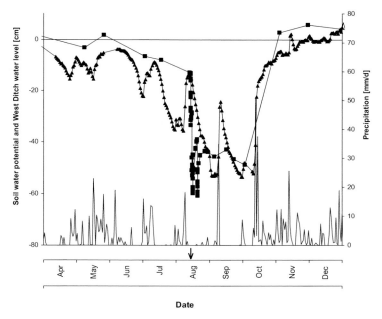

Figure 11. Relationship between precipitation, soil water potential (▲) and West Ditch water level (■) from April to December, 2002. The arrow indicates the onset of pump drainage.

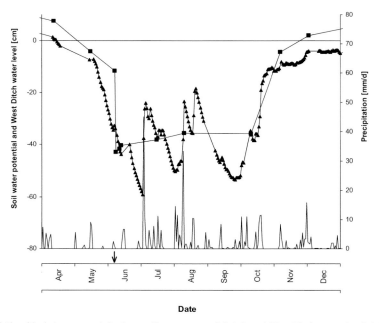

Figure 12. Relationship between precipitation, soil water potential (▲) and West Ditch water level (■) from April to December, 2001. The arrow indicates the onset of pump drainage.

Nutrient management in agricultural watersheds: A wetlands solution   **69**

in P sorption capacity, from 2032 mg kg^{-1} at 30 cm to 217 mg kg^{-1} at 90 cm. During only an eight day period of water table drawdown by intermittent pumping, as much as 400 g dissolved P (DP) entered the pumped ditch via subsurface flow from an eight ha area. Summer rainfall events greater than 35 mm d^{-1} also coincided with significant peaks in ditch water P concentration (up to 200 µg l^{-1} DP). Even larger peaks (up to 700 µg l^{-1} DP) occurred with the annual onset of autumn flooding. It is speculated that the permanently flooded, peat layer at around 90 cm depth probably provided a mass flow pathway for P to the pump drained ditch, owing to the low P affinity at this depth coupled with the high, lateral hydraulic conductivity. Moreover, redox profiles indicated that the dissolution of redox-sensitive Fe bound P may have contributed to the DP load at the 60-90-cm peat depth.

Acknowledgements

This work was financially supported by the European Union, Framework V (contract EVK1-CT1999-00036).

References

Faulkner, S.P., W.H. Patrick, and R.P. Gambrell, 1989. Field techniques for measuring wetland soil parameters. Soil Science Society of America Journal, **53** 883-890.

Fiedler, S., 2000. In situ long-term measurement of redox potential in redoximorphic soils. In Redox: Fundamentals, Processes and Applications edited by J. Schuring *et al.*, Springer-Verlag, Berlin, pp. 81-94.

Findlay, D.C., G.J.N Colborne, D.W. Cope, T.R. Harrod, D.V. Hogan, and S.J. Staines, 1984. Soils and their use in South West England. Soil Survey of England and Wales, Bulletin No. 14, Harpenden.

Heathwaite, A.L., 1991. Solute transfer form drained fen peat. Water, Air, and Soil Pollution **55** 379-395.

Hudson, J.A., and G. Roberts, 1982. The effect of a tile drain on the soil-moisture content of peat. Journal of Agricultural Engineering Research **27** 495-500.

Khalid, R.A., W.H. Patrick, and R.D. DeLaune, 1977. Phosphorus sorption characteristics of flooded soils, Soil Science Society of America Journal **41** 305-310.

Murphy, J., and J.P. Riley, 1962. A modified single solution method for the determination of phosphate in natural waters, Analytica Chimica Acta. **27** 31-36.

Nair, P.S., T.J. Logan, A.N. Sharpley, L.E. Sommers, M.A. Tabatabai, and T.L. Yuan, 1984. Interlaboratory comparison of a standardized phosphorus adsorption isotherm procedure. Journal of Environmental Quality **13** 591-595.

Nelson, D.W., and L.E. Sommers, 1982. Total carbon and organic matter. In: Methods of Soil Analyses, Part 2, edited by A.L. Page, R.H. Miller & D.R. Keeney, ASA. pp. 539-594.

Olsen, S.R., C.V. Cole, F.S. Watanabe, and L.A. Dean, 1954. Estimation of available phosphorus in soil extraction with sodium bicarbonate, USDA Circular 939. US Government Printing Office.

Rowell, D.L., 1994. Soil Science: Methods and Applications. Longman, Harlow, UK.

Scheffer, F., and P. Schachtschabel, 1998. Lehrbuch der Bodenkunde, 14th Edition, Enke Stuttgart, Germany.

Stewart, A.J.A., and A.N. Lance, 1983. Moor-draining - a review of impacts on land-use. Journal of Environmental Management **17** 81-99.

Stewart, A.J.A., and A.N. Lance, 1991. Effects of moor-draining on the hydrology and vegetation of northern pennine blanket bog. Journal of Applied Ecology **28** 1105-1117.

Syers, J.K., M.G. Browman, G.W. Smillie, and R.B. Corey, 1973. Phosphate sorption by soil evaluated by the Langmuir adsorption equation. Soil Science Society of America Proceedings **37** 358-363.

Watershed management and Reelfoot Lake: The role of wetlands

P.M. Gale
University of Tennessee at Martin, Department of Agriculture and Natural Resources, 256 Brehm Hall, Martin, TN 38238, USA

Abstract

Reelfoot Lake is considered one of the premier natural resources in West Tennessee and is the state of Tennessee's largest natural lake. Located on the floodplain of the Mississippi River, the lake is a popular destination for hunters and fishermen. Classified as a hypereutrophic system since the 1950's, the lake's water quality has been of concern to area resource managers. Although strides have been made in land use changes within the watershed, water quality in the lake has not improved. The objective of the current study was to evaluate the contribution of lake sediments to the water quality of Reelfoot Lake. Experiments were conducted to test the effects of a simulated drawdown on the release of N and P to the floodwater from the sediments. During a four week incubation, 12% of the total N and 2% of the total P were released from the sediments to the floodwater through the process of mineralization. These results demonstrate the potential contribution that sediments can make to the overall water quality of a lake.

Keywords: drawdown, nutrient mineralization, water quality.

Introduction

Reelfoot Lake is located on the Mississippi River floodplain in the northwest portion of the state of Tennessee. Uplift and subsidence in the region that occurred during the New Madrid earthquakes of 1811 and 1812 resulted in the formation of the lake (Johnston and Schweig, 1996). Prior to the earthquakes the area was floodplain forest crisscrossed by streams. Subsidence along the river channels of two of these streams resulted in the formation of the lake. Cypress trees are found throughout the lake today and delineate some of these old river channels, giving the lake the appearance of a flooded forest. The Reelfoot Lake watershed encompasses 62,200 ha and lies within two states. Lakes and wetlands across the Mississippi River floodplain have been found to be important carbon sinks (Delcourt and Harris, 1980).

The eastern portions of the watershed are silty uplands while the western portions are floodplain. Siltation in the lake has been recognized as a problem since the late 1800's. The surface area of the lake is 6200 ha and the average water column depth is 1.8 m. Sedimentation rates calculated using Cs profiles collected in 1984 were estimated at 1.5 cm yr^{-1} suggesting that the lifetime of the lake was around 120 years (Valentine *et al.*, 1994). Many of the streams in the uplands were channelised during the 1960's to enhance land drainage in this area. Channelisation along with increased crop production in the watershed greatly increased the sediment load into the lake (McIntyre and Naney, 1991).

The eastern portion of the watershed is comprised of highly erodible, silty uplands accounting for 93% of the sediment load into the lake. Estimated erosion rates in the watershed (based on land use) average 47 Mg ha^{-1} yr^{-1}. The maximum rate of 670 Mg ha^{-1} yr^{-1} was only estimated to occur on 50 ha (0.2% of the watershed) but could contribute a sediment load of 480 cm ha^{-1} to the lake (Denton and Dobbins, 1984). The majority of this sediment load, deposited in the wetlands that surround the lake, occurs from October to March. The sustainable erosion rate for these soils in this region is 11 Mg ha^{-1} yr^{-1} and conservation practices are being implemented in the watershed to meet this goal. The soils of the floodplain (western portions of the lakes watershed) are a mix of alluvial materials (sands, silts and clays) and contribute little in terms of sediment load but nutrient runoff from these agricultural fields is a concern. Forty seven percent of the watershed is in agricultural production. Row crops such as cotton, corn, rice and soybeans dominate the production.

The lake has been considered a hypereutrophic system since the 1950's and secchi depths of 5 to 10 cm are noted after storm events. According to the state's 2002 303(d) list 70.6% of the lake is listed as partially supporting while the remaining 29.4% is considered to be not supporting. In 1914 the state of Tennessee gained control of the lake from private landowners. A Watershed Plan was adopted by the Tennessee Department of Conservation and the Soil Conservation Service in 1959. This programme came into existence as the streams in the region were being channelized. In order to combat the extensive sediment loads from channelised streams in the watershed sediment control structures were constructed in the 1970's. These silt retention basins have been effective in the portions of the watershed that they serve. The state began baseline water quality monitoring of the lake in 1977 and in 1980 the Reelfoot watershed was selected as a national site for the Rural Clean Water Program. Through this programme a joint wastewater treatment facility was constructed for businesses and residences along the lake and the nearby town of Tiptonville. This drastically limited the use of onsite systems for wastewater treatment and it is estimated today that septic systems only contribute 0.1% of the nutrient load to the lake. The programme also encouraged the implementation of agricultural best management practices in the watershed including establishment of permanent cover, conservation tillage and enrollment of land into the conservation reserve program (Smith and Pitts, 1982).

The wetlands surrounding the lake are a mosaic of cypress - tupelo stands and open marsh vegetation. The marshes are dominated by giant cutgrass *Zizaniopsis miliacea Michx* while the littoral zones are home to spatterdock *Nuphar lutea L.* and coontail *Ceratophyllum demersum L.* Buildup of sediments in the open marsh areas have resulted in the establishment of water willow *Decodon verticillatus* and a subsequent loss of habitat for many wading bird species such as the Least Bittern. In the late 1980's the state introduced *Ctenopharyngodon idella* (grass carp) into the lake to control aquatic vegetation. Although they have become established *Ctenopharyngodon idella*, to date they have not had an anticipated negative effect on native fish species in the lake. However, many duck hunters suggest that since the establishment of grass carp the waterfowl habitat has been degraded especially through the loss of *Ceratophyllum demersum L.* one of the aquatic weeds the fish were introduced to control.

For several years lake managers have argued for a drawdown of the lake to consolidate the sediments and control aquatic vegetation. A drawdown may also have a significant impact on nutrient cycling in the lake. Previous studies in other systems have demonstrated movement of N and Fe with drawdown (Perrin *et al.*, 2000). Our objective was to examine the potential for movement of N and P from sediments of Reelfoot Lake under simulated drawdown conditions.

Materials and methods

In February of 2002, surficial sediment samples were collected using a ponar grab sampler, from 10 different locations around the lake. Samples were collected from areas where the water column was 0.5 m or less and represented areas that would be exposed during a drawdown. The locations were geo-referenced and distributed throughout the lake (Figure 1) to show each of the lake's wetland communities and basins. Samples were brought to the laboratory, dried (50°C, forced air oven) and ground (10 mesh, 2mm sieve) prior to analysis and experimentation.

For the incubation experiment, 15 cm^3 of dried and ground sediment material was placed into 100 cm^3 plastic bottles. This volume of sediment material provided weights ranging from 5 to 30 g, depending upon the bulk density of the material (Table 1). Three replicates of samples from each location on the lake were included. Fifty milliliters of filtered lake water was added to each bottle and the samples were capped, shaken and incubated in the dark at 25°C. Daily the sample bottles were opened for aeration, capped, shaken and returned to the incubator.

Floodwater samples from each bottle were collected once a week for a period of five weeks. For sample collection, 25 ml of the floodwater was removed, prior to shaking, and replaced with 25 ml of filtered lake water. The collected floodwater was filtered through 0.45 µm filters

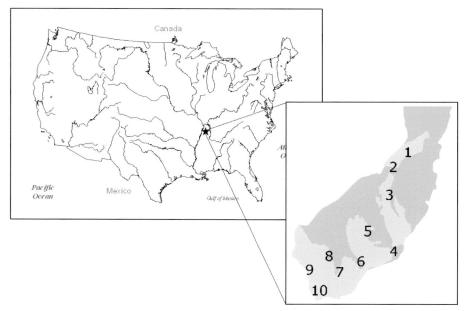

Figure 1. Location of Reelfoot Lake and the sampling stations used in the current study.

and analysed for NH_4^+, NO_3^- and soluble reactive phosphorus (SRP) using colorimetric techniques (APHA, 1992). Results were corrected for concentrations of these parameters in the floodwater and dilution during the experiment.

The remaining sediment material was sent to A&L laboratories (Memphis TN) for elemental analysis. Sediment analysis included digestion, dissolution and ICP analysis of the solution following procedures for solid wastes (USEPA, 1992). Total elemental analysis of the samples included N, P, K, S, B, Ca, Mg, Na, Fe, Al, Mn, Cu, Zn, Cd, Cr, Pb, Ni, Hg and As.

Results

Sediment characterisation data is presented in table 1. The bulk density of the sediment samples ranged from 0.27 to 1.99 g cm^{-3}. This variation in the composition of the sediment samples was also observed in other measured parameters and demonstrated a diversity of areas either dominated by organic or mineral substrates. Other studies have shown that accumulation of organic materials as sediments, can be a significant sink for carbon in the environment (Gale and Reddy, 1994).

The average percent organic matter of the samples was 17.4 while total N, P and S were 0.44, 0.13 and 0.16%, respectively. These data show that the sediments have a C:N:P ratio of 69:4:1, indicative of a material that would release mineral N and P during decomposition. Samples three and six, both high in organic matter and nutrients, represent areas that continue to receive agricultural runoff. The sediment samples having the lowest organic matter contents (seven, eight and nine) were located in the basin having the best water quality.

As the sediments decomposed during the incubation experiments N and P were released into the floodwaters. During the first four weeks of the incubation there was a net release of N and P to the floodwater, which subsequently levelled off. Nutrient release rates (Table 2) were calculated based upon the change in floodwater concentration with time during the four week period.

Table 1. Physico-chemical characterization of sediment samples collected for various locations on Reelfoot Lake.

Sample	Bulk density (g cm^{-3})	Organic matter (%)	Total N (%)	Total P (%)	Total S (%)
1	1.07	22.5	0.33	0.09	0.09
2	0.51	17.4	0.49	0.33	0.24
3	0.27	31.5	1.27	0.18	0.55
4	1.18	6.3	0.24	0.06	0.05
5	0.28	26.4	0.94	0.29	0.32
6	0.60	43.8	0.42	0.13	0.15
7	1.98	2.6	0.04	0.03	0.01
8	1.99	3.6	0.04	0.03	0.01
9	1.44	2.5	0.05	0.03	0.02
10	0.58	17.6	0.56	0.11	0.15

Table 2. Nutrient release rates and percent of total nitrogen and phosphorus mineralised during the incubation studies for sediment samples collected from Reelfoot Lake.

| Location | Nutrient Release Rate | | Nutrient Mineralised | |
	mg N kg^{-1} d^{-1}	mg P kg^{-1} d^{-1}	% total N	% total P
1	10.95	0.245	6.93	0.58
2	25.40	1.046	10.86	0.66
3	63.77	4.804	10.55	5.54
4	7.13	0.147	6.32	0.53
5	76.83	2.935	17.26	2.16
6	15.57	0.897	7.82	1.48
7	3.31	0.027	17.84	0.21
8	2.61	0.032	14.42	0.23
9	4.71	0.033	19.01	0.23
10	26.55	1.740	10.05	3.42

As expected, samples with high organic matter and low C:N ratios released higher amounts of nutrients to the floodwaters. Samples from sites 3, 5 and 10 released the highest amounts of nutrients and collectively had C:N ratios of 13:1. The lowest floodwater N and P concentrations were observed in samples from sites 7, 8 and 9 and these samples had an average C:N ratio of 31. Of the total N present in the sediments between 6 and 20% (average 12%) was mineralised during the four weeks of nutrient release. Likewise, the amount of total P mineralised ranged from 0.2 to 6% (average 2%). Overall the amount of N mineralized was 20 times greater than the amount of P mineralised.

Sediment metal concentrations (Table 3) varied also with organic matter and bulk density. Though the high bulk density samples were dominated by mineral materials, the organic matter

Table 3. Selected total metal concentration of sediment samples collected from Reelfoot Lake. All values are expressed on a dry weight basis.

| Location | % | | | mg kg^{-1} | | | |
	Mg	Ca	Fe	Al	Mn	Zn	Ni
1	0.174	0.607	1.228	4284	1149	34.5	10.4
2	0.442	0.722	3.751	17228	3207	121.6	26.2
3	0.405	0.730	3.121	14488	1352	105.6	25.7
4	0.208	0.762	0.821	3517	579	30.4	10.3
5	0.375	0.714	3.540	14284	1393	96.7	25.4
6	0.345	0.724	1.992	9191	1281	71.8	19.3
7	0.098	0.113	0.496	2176	352	15.2	5.5
8	0.108	0.123	0.515	1880	317	16.8	5.1
9	0.126	0.207	0.497	1878	605	15.5	5.7
10	0.387	0.716	1.499	8001	902	94.4	16.3

dominated materials had higher metal concentrations on a dry weight basis. The metal concentrations measured in this experiment fall within the range of data reported by Ammons *et al.* (1997). Their study characterised the total elemental composition of soil profiles sampled throughout Tennessee and included samples colleted from the Mississippi River valley silty uplands. These data provided further evidence that soil erosion in the surrounding watershed is contributing the mineral inputs into the lake.

Discussion

Our results suggest grouping the sediment samples into three categories based upon their bulk density and organic matter contents. However, these groupings were independent of location as each basin of the lake contained both high and low organic matter sediments along littoral zones (Figure 1).

The sediment samples could be broadly grouped into three categories that were shown to have similarities in their composition and reactivity. Samples one, three, five and six all had organic matter contents greater than 20%. Samples three and five had bulk densities less than 0.5 g cm^{-3}, high elemental concentrations and high nutrient release rates. Samples 2, 6 and 10 had bulk densities around 0.5 g cm^{-3}, high elemental concentrations and high nutrient release rates. While samples one, four, seven, eight, and nine had bulk densities greater than 1 g cm^{-3}, mineral soil properties (low nutrient concentrations) and low nutrient release rates. This grouping of sediments based upon organic matter and bulk density is evident in figures 2 and 3, where the amount of N mineralised is plotted versus these parameters. In figure 2 the highest organic matter samples (locations 3, 5 and 6) released the highest amounts of N during the study. Likewise, samples collected from locations three and five (bulk densities of 0.27 and 0.28 g cm^{-3}, respectively) also released the highest amounts of mineralized N (Figure 3).

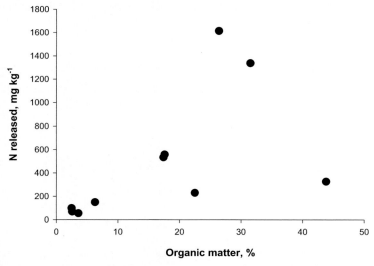

Figure 2. Relationship between N released during a five week incubation and the organic matter content of sediment samples collected from Reelfoot Lake.

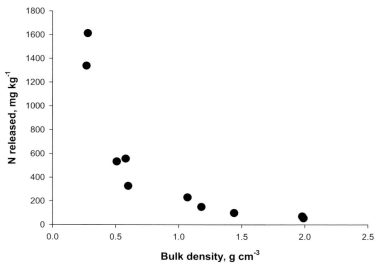

Figure 3. Relationship between N released during a five week incubation and the bulk density of sediment samples collected from Reelfoot Lake.

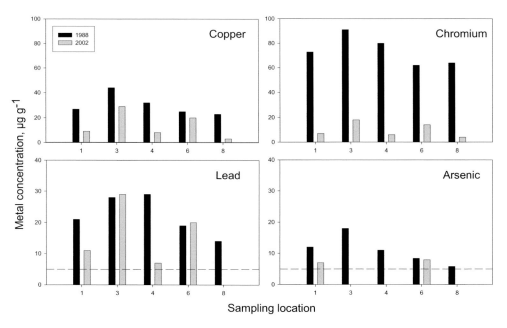

Figure 4. Changes in sediment metal concentrations between 1988 sampling (Broshears, 1991) and this study. Dashed line represents detection limit for analysis.

Five of the sampling locations used in this study were similar to a sediment study conducted by Broshears for the USGS (United States Geological Survey) in 1988 (Broshears, 1991). Figure 4 compares the data presented by Broshears with that generated during the current study. These graphs compare the concentrations of metals measured in the current study (gray bars) with those measured by Broshears in 1988 (black bars). Broshears (1991) concluded that the sediment metal concentrations were within the range of ambient levels for mineral soils of the region. The horizontal line depicts the detection limit for the metal. In general the concentrations of metals measured in the current study (2002 data) were lower than those measured in 1988 indicating a reduction in loading to the lake. For the most part, higher metal concentrations were associated with the higher organic matter sediments (samples three and six). The reduction in metal concentration of the low organic matter samples (from 1988 levels to 2002 levels) suggests burial of metals - inputs of these metals into the system have been curtailed during the past 20 years. In contrast, the sustained metal concentrations in the high organic matter sediments may suggest bioaccumulation and transfer of the metal by plants to the surface. In the 1988 sampling, Cd and Hg were detected in all of the samples; while in the current study concentrations of these metals were found to be below the detection limit (2.5 µg g^{-1} for Cd and 0.1 µg g^{-1} for Hg).

Conclusions

In 1987 the state listed and ranked the problems that managers of Reelfoot Lake face (Denton, 1987). At that time soil erosion and sedimentation were the main concern with water quality the secondary concern. Twenty years later this ranking has not changed, but today, managers have a better understanding of the importance of internal nutrient cycling of the system. Although Pardue *et al.* (1986) found no evidence of P migration into the water column from the sediments; our study has demonstrated that a drawdown of the lake could result in the release of significant amounts of soluble N and P to the water column. During the past 20 years no linkage has been made between land use and water quality in the lake (other than a noticeable decrease in the number and frequency of sediment plumes). Comparing current water quality data to STORET legacy data, there may be a slight decline in TKN levels (2.0 mg l^{-1} in 1984 and 0.9 mg l^{-1} in 1988) but TP has remained constant around 0.18 mg l^{-1} during this same time period. Hudson *et al.* (1999) have suggested that cycling of P within the water column is enough to sustain high planktonic populations. These data suggest that internal cycling of nutrients is significant in this ecosystem. The wetland areas of Reelfoot Lake provide accumulation areas for eroded materials from the surrounding watershed. Our results demonstrate that alternative management strategies (such as a drawdown) may change these nutrient sink areas into nutrient source areas.

Acknowledgements

Sample collection was conducted with the assistance of Reelfoot Lake State Park staff. Students Anthony Russo and Adam Hicks (Department of Agriculture and Natural Resources) provided assistance with sample collection and the incubation experiments. Funding for the elemental analysis of the sediment samples was provided by the Office of Research, Grants and Contracts, University of Tennessee at Martin.

References

American Public Health Association (APHA), 1992. Standard methods for the examination of water and wastewater. 18th Ed. American Public Health Association, Washington, DC.

Ammons, J.T., R.J. Lewis, J.L. Branson, M.E. Essington, A.O. Gallagher, and R.L. Livingston, 1997. Total elemental analysis from selected soil profiles in Tennessee. Tennessee Agricultural Experiment Station. Bulletin 693. Knoxville, Tennessee.

Broshears, R.E., 1991. Characterization of bottom-sediment, water and elutriate chemistry at selected stations on Reelfoot Lake, Tennessee. USGS Water Resources Investigations Report 90-4181.

Delcourt, H.R. and W.F. Harris, 1980. Carbon budget of the Southeastern U.S. biota: Analysis of historical change in trend from source to sink. Science **210** 321-323.

Denton, G.M., 1987. Water Quality at Reelfoot Lake - 1976 - 1986. Tennessee Department of Health and Environment. Nashville, TN

Denton, G.M. and W.R. Dobbins, 1984. A clean lakes study: The upper Buck Basin of Reelfoot Lake, Tennessee. Tennessee Department of Health and Environment, Division of Water Management. Contract ID-343109. Nashville, TN

Gale, P.M. and K.R. Reddy, 1994. Carbon flux between sediment and water column of a shallow, subtropical, hypereutrophic lake. Journal of Environmental Quality **23** 965-972.

Hudson, J.J., W.D. Taylor, and D.W. Schindler, 1999. Planktonic nutrient regeneration and cycling efficiency in temperate lakes. Nature **400** 659-661.

Johnston, A.C. and E.S. Schweig, 1996. The enigma of the New Madrid earthquakes of 1811-1812. Annual Reviews in Earth and Planetary Science **24** 339-384.

McIntyre, S.C. and J.W. Naney, 1991. Sediment deposition in a forested inland wetland with a steep-farmed watershed. Journal of Soil and Water Conservation **40** 64-66.

Pardue, J., D.H. Kesler, J. Dabezies, and L. Prufert, 1986. Phosphorus and chlorophyll degradation product profiles in sediment from Reelfoot Lake, Tennessee. Journal of the Tennessee Academy of Science **61** 46-49.

Perrin, C.J., K.I. Ashley, and G.A. Larkin, 2000. Effect of drawdown on ammonium and iron concentrations in a coastal mountain reservoir. Water Quality Research Journal of Canada **35** 231-244.

Smith, W.L. and T.D. Pitts, 1982. Reelfoot Lake: A Summary Report. Tennessee Department of Public Health, Division of Water Quality Control. Contract ID-1414, Nashville, TN

United States Environmental Protection Agency (USEPA), 1992. Test methods for Evaluating Solid Waste, Physical / Chemical Methods. US Environmental Protection Agency SW-846, 3rd edition.

Valentine, R., J. Mirecki, and E.S. Schweig, 1994. Estimate of historical sedimentation rate in a core from Reelfoot Lake, Tennessee. Journal of the Tennessee Academy of Science **69** 46.

Spatially distributed isolated wetlands for watershed-scale treatment of agricultural runoff

D.B. Perkins, A.E. Olsen and J.W. Jawitz
University of Florida, Soil and Water Science Department, 2169 McCarty Hall, Gainesville, Florida, 32611-0290, USA

Abstract

Non-point source phosphorus (P) loading from agricultural land use in the Lake Okeechobee Basin, Florida impacts the trophic level of the receiving rivers and Lake Okeechobee. To mitigate these impacts, the South Florida Water Management District has developed the Lake Okeechobee Action Plan. Part of this action plan includes the management of naturally occurring isolated wetlands (IWs), located throughout the landscape, to retain water and nutrients that may reduce the P loading from agricultural land to the lake. A simple landscape-scale one-dimensional steady-state water flow model was developed to analyze watershed P treatment of spatially distributed IWs in groundwater, applying a first-order decay rate model. Employing a range of model parameters typical of the Lake Okeechobee Basin, watershed-scale P treatment was compared for the following isolated wetland configurations: (1) a single isolated wetland, (2) a paired isolated wetland system with varying degrees of overlap in flow direction, and (3) a random field with five IWs. At relatively high decay rates, the level of P treatment decreased with the degree of isolated wetland overlap for the two-wetland configuration. For relatively low decay rates, P treatment increased with isolated wetland overlap. This analysis of P treatment using spatially distributed IWs in a landscape indicates that typical conditions of the Lake Okeechobee Basin are favourable for an effective means of reducing P from agriculturally impacted waters.

Keywords: isolated, wetlands, distributed, phosphorus, Okeechobee, treatment.

Introduction

Non-point source pollution, generated from leaching of nutrients and pesticides from agricultural land, is a critical environmental problem facing human society (Carpenter *et al.*, 1998; Reddy *et al.*, 2002; Tilman *et al.*, 2002). Non-point sources of agricultural pollution may cause a change in the trophic level of aquatic systems. This elevation in trophic level may result in reductions of water quality, loss of habitat and destruction of ecosystems. Eutrophication is a concern throughout the developed world, and specifically in the state of Florida where intensive agricultural practices generate large quantities of nutrient-laden water outflows. In particular, Lake Okeechobee has received phosphorus (P) loads that are considerably higher than historical averages over the last half-century (Bottcher *et al.*, 1995; Reddy *et al.*, 1999). Water quality has been adversely impacted within streams and rivers feeding the lake as well as within the lake itself.

In order to mitigate the negative impacts of agricultural practices within the Okeechobee Basin, the South Florida Water Management District has developed the Lake Okeechobee Action Plan

(Harvey and Havens, 1999). In this plan, several mechanisms were suggested for reducing P loading to the lake, including restoration of IWs within the basin by filling ditches that allow surface water to exit the isolated wetland systems. The restoration of IWs may result in the reduction of P loads to the lake via two mechanisms. The first is a reduction of total water reaching the lake. Water stored within IWs and the surrounding watershed will be lost to evapotranspiration (ET) thereby not entering the lake and not contributing to the P load. Secondly, the IWs will act as P filters within the watershed (Bottcher *et al.*, 1995; Harvey and Havens, 1999). By reducing P loads to Lake Okeechobee by restoration of isolated wetlands as a component of a larger mitigation effort, it is hoped that reductions in P loading to the lake be achieved and meet the legislative requirements as set out in the Lake Okeechobee Protection Act of 2000.

Literature documenting the definition, water quality, and geographic extent of IWs is summarised. Also, a water quality model was developed, incorporating typical parameters of the Lake Okeechobee Basin (a predominantly agricultural landscape), to evaluate the efficacy of using distributed IWs to reduce P loading under restored, non-ditched conditions.

Isolated wetland definitions

The term "isolated wetland" has been used to describe wetland systems that do not have surface water connectivity with rivers, lakes, oceans, or other water bodies (Leibowitz and Nadeau, 2003; Tiner, 2003a; Whigham and Jordan, 2003; Winter and LaBaugh, 2003). Soils, hydrology, and vegetation have been used to classify wetlands that interact directly with open water. The classification of IWs is less clear. The ambiguity associated with the classification of IWs is the result of an inconsistent definition that has been approached from hydrologic, ecologic, geographic, and/or other perspectives. For example, Whigham and Jordan (2003) defined an isolated wetland as strictly prohibiting any surface water connectivity with receiving water bodies at any time, while allowing for groundwater connections. Other researchers have allowed for infrequent surface water connectivity to nearby water bodies (Leibowitz and Nadeau, 2003; Tiner, 2003a; Winter and LaBaugh, 2003). Groundwater interaction with IWs is explicitly included in the definitions given by Snodgrass *et al.* (1996) and is recognized to play an important role in contaminant contribution to receiving water bodies. Ecologists have tended to define IWs by the occurrence of rare and highly dispersed habitats that are not adjacent to another body of water, which includes some notion of hydrology (NRC, 1995). Tiner (2003a) defined IWs in terms of geography, where the wetland must be completely surrounded by uplands consisting of non-hydric soils and non-aquatic vegetation.

Since some definitions may permit periodic surface water connectivity as well as considerable groundwater connectivity to receiving water bodies, the use of "isolated" is misleading from a hydrologic perspective. In reality, different degrees of surface and groundwater connectivity occur, and characterizing the degree of connectivity may be considered in evaluation of wetland interactions with distant receiving water bodies and ecological function. Winter and LaBaugh (2003) suggested the use of groundwater travel time as a parameter to establish the degree of hydrologic isolation, and as a standard to characterize a wetland as isolated. However, determining groundwater travel times for this purpose often requires rigorous and costly site characterization that may be impractical for large-scale classification of wetland connectivity to receiving water bodies. Therefore, to avoid the need to characterize highly variable hydrologic or ecologic processes, or to assess complex situations where some level of connectivity exists, the geographically-based definition proposed by Tiner (2003a) is used in this paper. Leibowitz and Nadeau (2003) also recommended this definition of isolation, which provides consistency for researchers while still using classical parameters with which most wetland scientists and regulators are familiar with.

Tiner (2003b) presented a comprehensive list of categories of IWs and their distribution across the contiguous United States and Alaska. To illustrate the possible range of IWs relative to the total amount of wetlands, a subset of data from a study completed by Tiner *et al.* (2002) that estimates the occurrence of geographically IWs is also reported here (Table 1). These data demonstrate that, regardless of climatic regime and areal coverage of wetlands in a landscape, IWs are of importance because of their potentially significant proportion of incidence.

Table 1 Estimated average percent of IWs by selected states as interpreted from Tiner et al. *(2002).*

State	Size of study area (ha)	% of study area in wetlands	Average % of wetland area predicted as isolated
FL	134,486	19	43.1
IL	119,776	5.4	59.5
KY	61,520	0.8	46.7
NJ	236,292	24.6	7.3
NE	173,460	2.7	51.0
NM	128,008	0.2	18.7
WA	104,427	1.6	78.3

Isolated wetlands water quality

Two factors may present challenges for characterizing water quality in IWs: (1) seasonal and hydrologic extremes alter the wetland's partitioning mechanism by either accreting or decomposing organic matter that in turn modify chemical release rates from the soil to the wetland water column, and (2) chemical inputs are diffuse non-point sources that may be spatially and temporally variable (Parsons *et al.*, 2003). While IWs are generally considered to be nutrient sinks (Hemond, 1980; Davis *et al.*, 1981; Ewel and Odum, 1984; Neely and Baker, 1989; and Pezzolesi *et al.*, 1998), it is difficult to determine whether a pattern exists for chemical transport to downstream water bodies. This is due to the paucity of investigations that adequately assess the impacts of hydrologic conditions.

In a review of water quality in IWs, Whigham and Jordan (2003) reported that the chemistry of surface waters in IWs is highly variable, being commonly linked to the input water source. The input water source may be heavily influenced by the watershed area and position of wetlands within a watershed. Smaller watersheds are conceptualized to contribute relatively lower chemical loads to the wetland from their uplands as compared to larger watersheds. For example, in an agricultural setting, nutrient and pesticide loading to an isolated wetland is directly proportional to the area of land that feeds it (Parsons *et al.*, 2003). According to a study conducted in prairie pothole wetlands, lower surface water salinity correlated to wetlands in higher topographic locations of a watershed (Driver and Penden, 1977). Surface water quality in IWs has also been correlated to the differences in underlying substrate thickness in the wetland and its chemical composition (Driver and Penden, 1977; Schalles, 1989; and Newman and Schalles, 1990).

Natural wetland systems have long been considered to be chemical buffers between uplands and receiving water bodies (Dortch, 1996; Reddy *et al.*, 1999; Price and Waddington, 2000). It was the recognition of such a function that spawned the idea of municipal storm-water and point-source agricultural effluent treatment using constructed wetlands (Gale *et al.*, 1994; Raisin *et al.*, 1997; Guardo, 1999; and Kovacic *et al.*, 2000). In the United States, treatment of such point sources is widespread, but implementing this concept for applications to non-point diffuse sources is less straightforward.

Isolated wetlands as treatment wetlands

In the case of an isolated wetland used for treatment of contaminants, two- or three-dimensional hydrologic and chemical monitoring may be necessary to estimate the treatment effectiveness and down-gradient chemical transport in surface and groundwater. A study by Shan *et al.* (2001) focused on an agricultural watershed in China that incorporated a network of spatially distributed mature constructed wetlands and ditches through which overland flow was routed before exiting the watershed. To handle diffuse contaminant loading, a calibrated curve number method was used to estimate runoff contribution to individual wetlands and the groundwater component was assumed to be negligible. This "multipond" system achieved greater than 90% retention for total phosphorus, dissolved phosphorus and suspended solids (Shan *et al.*, 2001). This study suggests that retention of contaminants from non-point sources may be greatly increased with a network of IWs; however, accurate estimation of diffuse contaminant loading to IWs is still challenging. Groundwater flows will provide an additional complicating factor for both water quality and quantity monitoring and system modeling. Additionally, seasonal variability in hydrologic and climatic conditions may complicate assessment of isolated wetland treatment efficiency (Parsons *et al.*, 2003).

Using a chloride tracer experiment in a naturally occurring isolated wetland in Canada, Hayashi *et al.* (1998) and Parsons *et al.* (2003) identified chemical cycling between the isolated wetland and the surrounding upland areas. In these studies, a barrage of water level and meteorological instrumentation was used to characterize groundwater flow pathways and water balance inputs. Lateral groundwater flow, evapotranspiration, and deep aquifer recharge were the only hydrologic components that occurred in this isolated wetland. It was completely severed from any surface water interactions with adjacent water bodies and, while not specifically commented on by the authors, the down-gradient export of contaminants would likely change if the terrain were modified by ditching. Compared to contaminant loading in groundwater outflow, ditching may allow greater export of contaminants from the isolated wetland due to relatively faster hydrologic response time and reduced mean residence time within the wetland.

The findings from these investigators lend confidence to the idea that IWs may be effective for contaminant removal from diffuse groundwater and surface water sources. Also, the degree of surface water connectivity and water movement to a nearby water body plays an important role in the removal rates of contaminant inputs because mean residence time (τ) generally decreases as water velocity increases (Mitch and Gosselink, 2000):

$$\tau = \frac{V \varepsilon}{q A} \qquad (1)$$

where V is the volume of water involved in the wetland (l^3), ε is the water fraction in the flow media, q is the specific discharge ($l\ t^{-1}$), and A is the cross-sectional area of flow (l^2).

The objective of this study was to demonstrate the efficacy of using spatially distributed IWs for the treatment of P-contaminated water within an agricultural watershed. A first-order model was developed that simulated water treatment within a watershed similar to those located in the Okeechobee Basin. Simulations were performed at various levels of wetland percent coverage, spatial distribution of wetlands, and contaminant degradation rate. These simulations show that spatially distributed wetland may provide an effective means of treating contaminated water within a watershed.

Isolated treatment wetlands model

The efficacy of using spatially distributed IWs for the treatment of non-point sources of contamination within a watershed was evaluated using a first-order contaminant degradation model. Specific application of first-order decay kinetics to treatment wetlands has its origins in work described by Kadlec and Knight (1996). Water flow was assumed to be steady-state. The watershed was divided into a regular grid that was aligned with the mean flow direction. Each cell within the grid was treated as a completely-stirred treatment reactor (CSTR). This discretization of the watershed also assumed that lateral mixing was minimal, such that water flowed through a single column in the grid and did not mix between columns. The solute mass balance may be expressed as shown in Equation 2:

$$\frac{V_w\, d\, C}{d\, t} = V_w C_{in} G_{in} - V_w C_{out} G_{out} + V_w C_a A - V_w k C \qquad (2)$$

where C is the solute concentration within a grid cell (M l^{-3}), V_w is the volume of a grid cell (l^3), C_{in} is the solute concentration entering a grid cell via groundwater inflow (m l^{-3}), C_{out} was the solute concentration leaving a grid cell via groundwater outflow (m l^{-3}), C_a is the areal contaminant addition rate (m $l^{-3}\ l^{-2}\ T^{-1}$), and k is the volumetric decay coefficient (T^{-1}). Assuming that each grid cell acts as CSTR, the solute concentration as a function of time may be written as

$$C(t) = C_{in} e^{-kt} + \frac{C_a A}{k} (1 - e^{-kt}) \qquad (3)$$

These assumptions result in a tanks-in-series model and the exact solution of Equation 3 can be determined by recursion.

Because the degradation properties of the wetland and non-wetland portions of the watershed were different it was necessary to determine what proportion of each cell was composed of wetland. This requires wetlands of known geometry and location. To simplify the model it was assumed that each wetland was shaped like the bottom portion of a sphere (circular in plan view and semi-circular in cross section). A stage-surface area relationship was then developed so that the surface area of the wetland within the watershed could be determined for any wetland depth

$$A_w = 2\pi \, h d_w \tag{4}$$

where A_w is the surface area of the wetland [l^3], h is the radius of the sphere [l], and d_w is the depth of the wetland [l]. The semi-circular assumption also allowed for the development of a stage-volume relationship

$$V_w = \left(\tfrac{2}{3}\pi h^2 \, d_w\right) - \left(\tfrac{1}{3}\pi r_w^2 \, (h - d_w)\right) \tag{5}$$

where V_w is the volume of the wetland [l^3].
When the total wetland volume was determined using the model it agreed closely with theoretical wetland volumes.

Three scenarios were considered in a landscape of constant area (10^6 m^2 (10^3 by 10^3 m)): (1) a single wetland, (2) two wetlands, and (3) a random distribution of five IWs. The total wetland area was the same for all scenarios and was constrained to a maximum of 16% of the watershed area, based on the average isolated wetland areal coverage in the Lake Okeechobee Basin, FL (Flaig and Reddy, 1995). From estimated cow grazing densities typical of the Okeechobee Basin (0.3 cows/acre), a constant P input of 0.891 mg l^{-1} was applied in all simulations. The single-wetland simulation scheme was implemented to identify the degree of total water outflow treatment in the simplest case. The center of the isolated wetland was placed in the geographic center of the landscape (Figure 1A). This scenario also served as a benchmark for comparing multiple-wetland scenarios of varying spatial arrangements.

The two-wetland arrangement was implemented to explore groundwater treatment of various spatial arrangements of IWs in the landscape. In this one-dimensional approach, water treated from an isolated wetland may, to varying degrees, contribute to the inflow of an isolated wetland that is further down the flow path. This contribution is termed isolated wetland overlap and affects the total watershed P treatment. For all simulations, the water flow was in the y-direction (Figure 1). In the first of four simulations of two-wetland arrangements, the IWs were offset in the y-direction to avoid overlapping of individual wetland areas and were just far enough apart in the x-direction so that no treated water from the first sequential wetland would enter the second. Preserving the offset in the y-direction in the following four simulations, the wetlands were systematically moved closer together in the x-direction to allow more treatment overlap in the flow-field with each simulation until the water flow path of the second wetland eclipsed the first (Figure 1b-e).

One simulation of five randomly distribution IWs was performed, maintaining the same maximum areal coverage of 16% (Figure 1f).

Treatment efficiency (ξ) was used to evaluate watershed-scale P retention:

$$\xi = \left[1 - \left(\frac{c}{c_0}\right) * 100\right] \tag{6}$$

where c is the total P concentration along the watershed outflow boundary (m l^{-3}) and c_0 is the total concentration added to the watershed for the duration of the simulation (m l^{-3}). The term c/c_0 is the residual normalised concentration at the watershed outflow boundary.

Two model parameters were varied with each simulation: (1) wetland area (A_w), calculated as a function of depth of water in the wetland(s), and (2) decay rate (k_v). The wetland area was incrementally increased from 0.01% of the total watershed area to 16% by increasing the level of water in the wetland from 0 to 1 m depth. This depth is consistent with typical depths of IWs in the Lake Okeechobee Basin. The k_v was varied from 0.1 to 10 yr^{-1} for each incremental increase in A_w. Gale *et al.* (1994) reported k_v values for the constructed wetland treatment

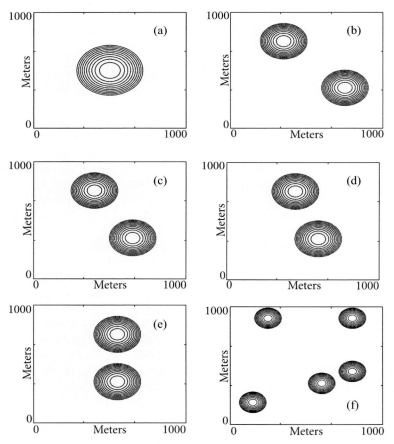

Figure 1. (a) single IW in the center of the watershed, (b-e) two-IW system with increasing overlap in the flow direction, and (f) random IW system.

system in Orlando, FL between 12.8 and 26.3 yr^{-1}. Sompongse (1982) found k_v values ranging between 18 and 29 yr^{-1} for Florida soils receiving agricultural drainage waters. A somewhat lower range of k_v was used relative to the literature values for Florida systems in an effort to offset the slow groundwater component and simulate a system in relative non-equilibrium. The degree of non-equilibrium was characterized using the Damkohler number (D_a):

$$D_a = k_{v,w} \, \tau_w \tag{7}$$

where $k_{v,w}$ is the decay coefficient for a wetland and τ_w is the mean residence time.

Results and discussion

Treatment efficiency (ξ)

The high end range of k_v values used in the simulations in this study is similar to those reported by other investigators (Sompongse, 1982; Gale *et al.*, 1994). These relatively higher k_v values resulted in equilibrium decay conditions that may not reflect rate-limited P dynamics that are usually manifested in natural systems. Therefore, the low range of k_v values were included in the simulations to create more rate-limited P decay, which placed more emphasis on the hydraulic contact area of the wetlands. Increased wetland hydraulic contact area was generally correlated to higher P treatment.

The ξ for the single isolated wetland simulation (first scenario) ranged from 0 to 20% of the annual P addition. While the ξ, k_v, and fractional wetland area (A_w) appear exponentially related, the A_w and k_v are of near-equal influence for watershed-scale groundwater P treatment (Figure 2a). This type of relationship between ξ, A_w, and k_v persists for all simulations of different numbers and arrangements of IWs, but the relative range of treatment varies slightly.

For the second scenario of four simulations of two-wetland arrangements, the arrangement having the least amount of overlap in the flow direction yielded the highest maximum P treatment of 28% (Figure 2b). The other three wetland arrangements, with incrementally increasing overlap, resulted in increasingly smaller maximum P treatments (27, 25, and 20%, respectively) (Figure 2c, 2d, and 2e).

The final scenario of five randomly distributed isolated wetlands estimated maximum P treatment across the outflow boundary to be 28% (Figure 2f). This P reduction was similar to the scenario made up of two wetlands that incorporated the least overlap in the flow direction because of increased interception of flow through the landscape.

Analysis of the two-wetland arrangements (Figure 3) showed that at $k_v = 0.1$, the length term embedded in τ_w (Equation 1), representing the water flow path through wetland grids, dominated the D_a. This resulted in an increase in treatment efficiency with increasing fraction of overlap of the two wetlands. The D_a generated from $k_v = 0.1$ with the maximum and minimum wetland areal coverage was 23 and 13, respectively, and represented an equilibriu0m P decay processes. Equilibrium decay conditions were also observed ($D_a > 100$) when k_v approached 10. In this case, maximum treatment is observed at zero overlap.

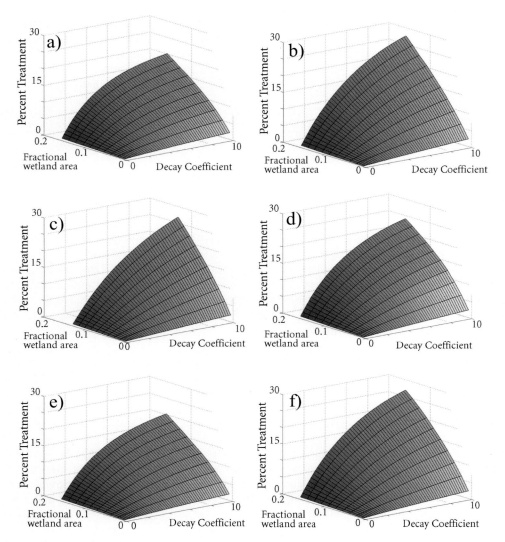

Figure 2. Treatment efficiency for (a) the single isolated wetland system, (b-e) the two-IW system with increasing overlap, and (f) the random IW distribution. Treatment efficiency is plotted as a function of fractional wetland areal coverage and decay coefficient (k_v).

Effect of wetland overlap

Evaluation of wetland overlap and P treatment at high and low k_v values adds insight to the importance of the relative geographic position of IWs in the landscape. Wetland overlap was defined as the radial overlap of the two wetlands in the flow direction divided by the wetland diameter (as both wetlands were the same size). Simulations using relatively lower k_v values

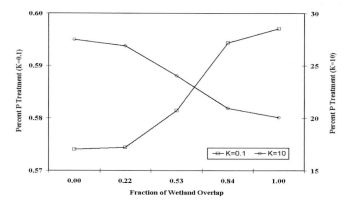

Figure 3. Fractional overlap for a two-wetland system, with relatively high and low k_v values of 10 and 0.1, respectively.

showed that while the treatment difference between scenarios of complete or no wetland overlap was small (~4%), more wetland overlap resulted in better treatment because water treatment flow path controls the D_a. Simulations involving relatively higher k_v values revealed higher P treatment when the two wetlands were not overlapping in the flow direction. In these scenarios, corresponding to near- instantaneous P decay, the change in length of the wetland flow path was not great enough to produce differences in treatment from a single-wetland scenario; rather it was more important to have more of the water that passed through the landscape treated.

Conclusions

To date, no definitions of IWs have been promulgated by regulatory agencies. Therefore, various definitions have emerged in the literature, and these depend on the discipline of origin and the intended application. The geographic definition offered by Tiner (2003a) was used here because it provides consistency for regulators and scientists, while applying commonly used soil and vegetation parameters without the need to intensively characterise watershed parameters.

Modeling P reduction in an isolated wetland landscape provided insight into the utility of such a landscape for treatment of agriculturally impacted groundwater. In a single isolated wetland system, at the maximum areal coverage and with a relatively high k_v, reduction of P in groundwater discharge along the outflow boundary was approximately 20%. Reduction of P in the landscape increased to 28% for simulation of a system of two IWs with the same fractional area of IWs that did not overlap in the direction of flow. But, increased overlap in the flow direction corresponded to decreased P reduction.

In conjunction with other proposed land management practices and P reduction strategies proposed in the Okeechobee Action Plan (Harvey and Havens, 1999), the distributed IW component analysed in this study could contribute to P load reductions from agriculturally dominated landscapes. The model developed here indicates that, under restored conditions (non-ditched), spatially distributed IWs in the Lake Okeechobee Basin may considerably reduce P loading to receiving water bodies.

Acknowledgements

Support for project funding was provided by the Florida Department of Agricultural and Consumer Services, the South Florida Water Management District, and the Florida Department of Environmental Protection. Primary funding for the first two authors was provided by the Florida Alumni Graduate Fellowship. The authors would also like to acknowledge the assistance of Dayna Perkins and Julie Padowski for their valuable contributions.

References

Bottcher, A.B., T.K. Tremwel, and K.L. Campbell, 1995. Best management practices for water quality improvement in the Lake Okeechobee Watershed. Ecological Engineering **5** 341-356.

Carpenter, S.R., N.F. Caraco, D. L. Correll, R.W. Howarth, A.N. Sharpley, V.H. Smith, 1998. Nonpoint pollution of surface water with phosphorus and nitrogen. Ecological Applications **8** 559-568.

Davis, C.B., J.L. Baker, A.G. van der Valk, and C.E. Beer, 1981. Prairie pothole marshes as traps for nitrogen and phosphorus in agricultural runoff. pp. 153-63. In: Riparian Management in Forests of the Continental Eastern United States, edited by B. Richardson, Lewis Publishers, Boca Raton, FL, USA.

Dortch, M.S., 1996. Removal of solids, nitrogen, and phosphorus in the Cache River Wetland. Wetlands **16** 358-365.

Driver, E.A. and D.G. Peden, 1977. The chemistry of surface water in Prairie ponds. Hydrobiologia **53** 33-48.

Ewel, K.C. and T.T. Odum (eds.)., 1984. Cypress Swamps. University Presses of Florida, Gainesville, FL, USA.

Flaig, E.C. and K.R. Reddy, 1995. Fate of phosphorus in the Lake Okeechobee watershed, Florida, USA. Ecological Engineering **5** 163-181.

Gale, P.M., K.R. Reddy, D.A. Graetz, 1994. Phosphorus retention by wetland soils used for treated wastewater disposal. Journal of Environmental Quality **23** 370-377.

Guardo, M., 1999. Hydrologic balance for a subtropical treatment wetland constructed for nutrient removal. Ecological Engineering **12** 315-337.

Harvey, R. and K. Havens, 1999. Lake Okeechobee Action Plan. South Florida Water Management District. West Palm Beach, FL. pp. 1-43.

Hayashi, M., G. van der Kamp, and D.L. Rudolph, 1998. Water and solute transfer between a prairie wetland and adjacent uplands, 1. Water balance. Journal of Hydrology **207** 42-55.

Hermond, H.F., 1980. Biogeochemistry of Thoreau's Bog, Concord, MA. Ecological Monographs **50** 507-526.

Kovacic, D.A., M.B. David, L.E. Gentry, K.M. Starks, and R.A. Cooke, 2000. Effectiveness of constructed wetlands in reducing nitrogen and phosphorus export from agricultural tile drainage. Journal of Environmental Quality **29** 1262-1274.

Leibowitz, S.G. and T. Nadeau, 2003. Isolated wetlands: state-of-the-science and future direction. Wetlands **23** 663-684.

Mitch, W.J. and J.G. Gosselink, 2000. Wetlands (Third Edition). John Wiley, New York, NY, USA.

National Research Council (NRC), 1995. Wetlands: Characteristics and boundaries. National Research Council Committee on Characterization of Wetlands, National Academy Press, Washington DC, USA.

Neely, R.K. and J.L. Baker, 1989. Nitrogen and phosphorus dynamics and the fate off agricultural runoff. pp. 92-131. In: Northern Prairie Wetlands, edited by A.G. van der Valk. Iowa State University Press, Ames, IA, USA.

Newman, M.C. and J.F. Schalles, 1990. The water chemistry of Carolina bays: A regional survey. Archiv Für Hydrobiologie **118** 147-168.

Parsons, D.F., M. Hayashi, and G. van der Kamp, 2003. Infiltration and solute transport under a seasonal wetland: bromide tracer experiments in Saskatoon, Canada. Hydrological Processes **18** 2001-2027.

Pezzolesi, T.P., R.E. Zartman, E.B. Fish, and M.G. Hickey, 1998. Nutrients in Playa wetland receiving wastewater. Journal Environmental Quality **27** 67-74.

Price, J.S. and J.M. Waddington, 2000. Advances in Canadian wetland hydrology and biogeochemistry. Hydrological Processes **14** 1579-1589.

Raisin, G.W., D.S. Mitchell, and R.L. Croome, 1997. The effectiveness of a small constructed wetland in ameliorating diffuse nutrient loading from an Australian rural catchment. Ecological Engineering **9** 19-35.

Reddy, K.R., G.A. O'Connor, and C.L. Schelske. (eds.)., 1999. Phosphorous biogeochemistry in subtropical ecosystems. Lewis Publishers Inc., Boca Raton, Florida, 707 pp

Reddy, K.R., O.A. Diaz, L.J. Scinto, and M. Agami, 1995. Phosphorus dynamics in selected wetlands and streams of the lake Okeechobee Basin. Ecological Engineering **5** 183-207.

Schalles, J.F., 1989. Comparative chemical limnology of Carolina bay wetlands on the Upper Coastal Plain of South Carolina. pp. 89-111. In: Freshwater Wetlands and Wildlife, edited by R.R. Sharitz and J.W. Gibbons, USDOE Office of Science and Technical Information, Oak Ridge, TN, USA.

Shan, B., C. Yin, and G. Li., 2002. Transport and retention of phosphorus pollutants in the landscape with a traditional, multipond system. Water, Air, and Soil Pollution **139** 15-34.

Snodgrass, J.W., A.L. Bryan, Jr., R.F. Lide, and G.M. Smith, 1996. Factors affecting the occurrence and structure of fish assemblages in isolated wetlands of the upper coastal plain, U.S.A. Canadian Journal of Fisheries and Aquatic Sciences **53** 443-454.

Sompongse, D., 1982. The role of wetland soils in nitrogen and phosphorus removal from agricultural drainage water. Ph.D. Diss. Univ. of Florida, Gainesville (Diss. Abstr. DA 8226434).

Tilman, D., K.G. Cassman, P.A. Matson, R. Maylor, S Polasky, 2002. Agricultural sustainability and intensive production practices. Nature **418** 671-677.

Tiner, R.W., H.C. Bergquist, G.P. DeAlessio, and M.J. Starr, 2002. Geographically isolated wetlands: a preliminary assessment of their characteristics and status in selected areas of the Unites States. U.S. Department of the Interior, Fish and Wildlife Service, Northeast Region, Hadley, MA, USA (web-based report at: wetlands.fws.gov).

Tiner, R.W., 2003a. Geographically isolated wetlands of the United States. Wetlands **23** 494-516.

Tiner, R.W., 2003b. Estimated extent of geographically isolated wetlands in selected areas of the United States. Wetlands **23** 636-652.

Whigham, D.F. and T.E. Jordan, 2003. Isolated wetlands and water quality. Wetlands **23** 541-549.

Winter, T.C. and J.W. LaBaugh, 2003. Hydrologic considerations in defining isolated wetlands. Wetlands **23** 532-540.

Nitrogen cycling in wetland systems

P.G. Hunt, M.E. Poach and S.K. Liehr
USDA-ARS, 2611 W. Lucas St., Florence, SC, 29501, USA

Abstract

When considering the management of N on an agricultural watershed, the cycling of N is paramount because N exists in many different oxidative and physical states. The cycle is active in the biology of both aerobic and anaerobic processes. Furthermore, the cycling of N in both natural and constructed wetlands is particularly dynamic and exceedingly valuable to N management for both productive agriculture and environmental quality. In this chapter, we illustrate N cycling in the context of three types of wetlands - constructed, riparian, and in-stream. We present the higher than expected rates of denitrification in constructed wetland used for animal wastewater treatment as an example of denitrification via new pathways such as ANAMMOX that require less oxygen in the precursor oxidation of ammonia. We show the effectiveness of different riparian zones for stream buffering and denitrification, particularly noting that they appear to provide a reasonable balance for protecting both water and air quality. We emphasize the importance of in-stream wetlands for assimilations and transformations of N that escapes agricultural watersheds.

Keywords: ammonia, nitrification, denitrification, riparian, in-stream.

Introduction

The topic of recycling N in wetland systems as it relates to nutrient management in agricultural watersheds has a long and interesting history. The topic can be discussed from the watershed system to the molecular level. In this chapter, we will discuss some aspects of both the watershed and microbial scales. Our approach will be to present N cycling in three wetland components of a typical agricultural watershed to illustrate how wetland systems can be used to manage nutrient losses. The wetland components will be: 1) constructed wetland for treating animal wastewater, 2) riparian zones for buffering N losses from adjoining fields, and 3) in-stream wetlands for removing N that has entered drainage ditches and streams.

Constructed wetlands

Treatment of animal wastewater in constructed wetlands in the USA has drawn heavily from the experience and technology derived from early research on agricultural wetland systems and on treating municipal wastewaters in natural systems (Patrick and Mikkelsen, 1971; Hunt and Lee, 1975 and 1976; Hunt *et al.*, 2002; Kadlec and Knight, 1996; Reed *et al.*, 1995). Sadly, two of these early researchers Woody Reed and Bill Patrick have passed away recently. Patrick's early research documented the importance of the close juxtaposition of the aerobic and anaerobic layers of the soil for N cycling in rice production (Patrick and Mikkelsen, 1971). The connection between the nitrogen cycle of wetland agriculture (i.e., rice) and wastewater treatment was made by Hunt and Lee (1975 and 1976). They, in cooperation with the Campbell Soup Company, documented how the N cycled in the surface of an overland flow treatment system to allow the N contained in food canneries or municipalities wastewaters to be simultaneously mineralized, nitrified, and denitrified (Hunt *et al.*, 1976). They also documented how these concepts could be used in conjunction with wetland vegetation to improve water quality of effluent from dredged materials that were deposited in constructed land containment areas near rivers and estuaries (Lee *et al.*, 1976). The ideas of Siedel (1976) in Europe were proceeding along similar lines. Additionally, Maxwell Small (Small and Wurm, 1977) was exploring the use of wetlands to treat municipal wastewaters in the USA. Later, excellent research and writing by Hammer, Reed, Crites, Kadlec, Knight, Mitch, and others (Hammer, 1989; Kadlec and Knight, 1996; Reed *et al.*, 1995; Mitch and Gosselink, 2000) provided a very good understanding of wetland function and design, including N cycling.

In the USA, the Natural Resources Conservation Service of the US Department of Agriculture established an Engineering Standard for constructed wetlands, which was based on organic loadings measured as (biological oxygen demand) BOD, but it was directed toward assisting livestock producers in meeting the limits on land application of N from livestock wastewaters (USDA, 1991; Stone *et al.*, 2002). Several investigations documented the large N removal capacities of constructed wetlands, even when loaded at rates greater than 20 kg ha^{-1} d^{-1} (Cathcart *et al.*, 1994; McCaskey *et al.*, 1994; Knight *et al.*, 2000). While soil adsorption and plant recycling were significant components of N cycling when the constructed wetlands received relatively low loading of N (Szogi *et al.*, 2000), they became a minor portion of the annual budget when loads exceeded 10 kg ha^{-1} d^{-1} (Knight *et al.*, 2000; Hunt *et al.*, 2002). Thus, the N removals were deemed to be gaseous by either direct ammonia volatilization or denitrification. Poach *et al.* (2002) conducted investigations to document the extent of ammonia volatilization from constructed wetlands loaded with swine effluent from anaerobic lagoons. He found that volatilization was quite significant over the open water areas of marsh-pond-marsh wetlands when these wetlands were loaded at rates of 12 kg ha^{-1} d^{-1} or more (Poach *et al.*, 2002 and 2004; Figure 1). However, ammonia volatilization was a minor component of the N losses when the wetlands had good vegetative cover. Furthermore, he showed that even partial pre-wetland nitrification would significantly lower ammonia volatilization losses (Poach *et al.*, 2003).

These findings led to the conclusion that very significant denitrification was occurring in the constructed wetland. This high level of denitrification was corroborated by the levels of denitrification enzyme activity (DEA) in constructed wetland systems (Hunt *et al.*, 2003). They found that the three-year DEA mean in the soil surface was equivalent to 9.55 kg N ha^{-1} d^{-1} in wetlands dominated by bulruhes, and the denitrification potential from the surface litter and floating sludge matrix were much higher (Hunt *et al.*, 2002a; Hunt *et al.*, 2002b).

While these high levels of denitrification were encouraging, they raised questions about oxygen for the precursor nitrification. Transformation of N in constructed wetlands is generally considered to be oxygen limited (Patrick and Reddy, 1976; Reed, 1993) and this limitation would only be exacerbated by the organic- and ammonia-N levels in high-strength wastewaters. Moreover, nitrate rather than carbon was found to be the clear limiting factor for denitrification in constructed wetlands treating swine wastewater (Hunt *et al.*, 2003). The O_2 that is assumed to be necessary for nitrification is not typically measured in these treatment systems. However, this oxygen limitation appears to be less fixed than once thought. Tanner and Kadlec (2003)

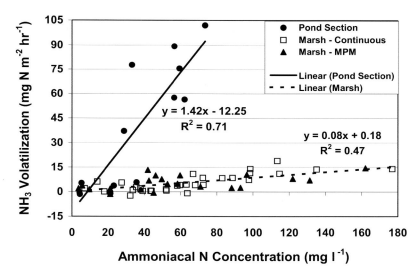

Figure 1. Ammonia volatilization from constructed wetlands used to treat swine wastewater (Poach et al., 2004).

concluded that significantly more N losses were occurring in constructed wetlands than could be explained by the apparent oxygen transfer. Fortunately, there have been significant advances in the understanding of N cycling that offer some reasonable explanations for these high levels of denitrification (Figures 2 and 3).

These observations have led to re-examination of the issue of N transformation in animal waste systems. Studies of N conversion under low-oxygen and anaerobic conditions have shown that

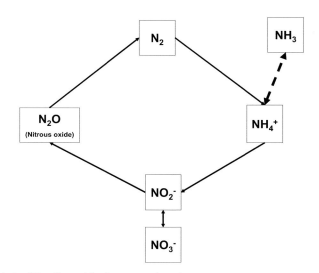

Figure 2. Basic N cycle traditionally used in the agronomic and wastewater treatment context.

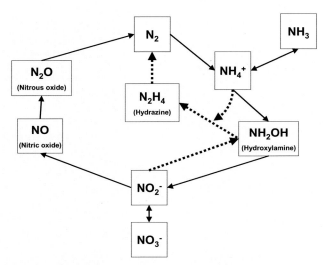

Figure 3. Nitrogen cycle including alternate pathways for denitrification (Modified from Jetten, 2001).

ammonia can be converted to di-nitrogen by processes other than conventional nitrification of ammonia to nitrate followed by denitrification of nitrate to di-nitrogen gas. Under low-oxygen conditions, the production of nitrite from ammonia is favored over the production of nitrate (Bernet *et al.*, 2001; Hanaki *et al.*, 1990). The nitrite can then be denitrified to nitrous oxide and/or di-nitrogen without first being converted to nitrate. This process has been termed 'partial nitrification-denitrification.' Under low-oxygen conditions, ammonia oxidisers were able to denitrify the nitrite, which they produced from the conversion of ammonia to nitrous oxide and di-nitrogen (Poth and Focht, 1985; Poth, 1986; Brock *et al.*, 1995; Muller *et al.*, 1995). Whereas one organism was able to perform both conversions, this process has been termed 'combined nitrification-denitrification' (Kuenen and Robertson, 1994). Ammonia oxidisers were also able to convert ammonia to di-nitrogen with N dioxide as the oxidant under anaerobic conditions (Brock *et al.*, 1995; Schmidt and Bock, 1997). Other organisms can also oxidise ammonia to di-nitrogen under anaerobic conditions using hydroxylamine derived from nitrite as the oxidant. This process has been termed 'anaerobic ammonia oxidation' or ANAMMOX (Mulder *et al.*, 1995).

Typically, the assumption has been that ammonia oxidation requires high concentrations of O_2 (>1.0 mg l^{-1}). However, recent research has shown that this assumption is incorrect. Observations have been made in wastewater treatment systems where aerobic ammonia oxidation occurred at DO concentrations less than 0.4 mg l^{-1}, and in some cases even at less than 0.1 mg l^{-1} (Park *et al.*, 2002; Sliekers *et al.*, 2002; and Bernet *et al.*, 2001). Therefore, the N transformation capacity depends more on the rate of oxygen transfer than on measured oxygen concentrations.

Most of the constructed wetlands built in the USA for treatment of swine and dairy effluents are surface flow (FWS) wetlands. For these types of wetlands, the major oxygen transfer site is the water surface. Factors that affect the transfer rate include temperature, dissolved solids, dissolved oxygen concentration, wind and water currents that promote mixing, and other physical forces that cause mixing. Estimates of transfer rates have been made based on techniques used with lakes, ponds, and streams. These methods do not take into account the effect of emergent vegetation on the transfer processes; and because of the many factors affecting transfer, there is still a high level of uncertainty in these estimates. Kadlec and Knight (1996) used lake transfer models (USEPA, 1985) to estimate transfer rates of O_2 for a range of wind speeds. At 20°C, the range of transfer rates of O_2 for wind speeds of 1-10 m sec^{-1} was 4.5 to 450 kg ha^{-1} d^{-1}. The lake environment more closely resembles a free water surface wetland, and these numbers are likely more representative than estimates of 12 - 30 kg ha^{-1} d^{-1} made using stream models (Kadlec and Knight, 1996; O'Connor and Dobbins, 1958; Mattingly, 1977). A typical transfer rate for average wind speeds would be in the range of 30 - 45 kg ha^{-1} d^{-1}. However, actual transfer rates may be more or less than this estimate due to a number of factors that contribute to the uncertainty, including effect of emergent vegetation, rainfall, recycle practices, inlet structures, plant root transfer, and inputs from algae and submersed aquatic vegetation.

How much oxygen would it take to remove a typical amount of ammonia N from animal wastewater? For standard nitrification / denitrification

Nitrification: $NH_4^+ \Rightarrow NO_2^- \Rightarrow NO_3^-$

Denitrification: $NO_3^- \Rightarrow NO_2^- \Rightarrow N_2$

4.6 g of O_2 are required for 1.0 g of NH_4-N. This means that for a typical N loading rate of 20 kg ha^{-1} d^{-1}, 92 kg ha^{-1} d^{-1} of O_2 would be required. Although this is within the larger range including higher wind speeds, it is somewhat higher than the range that might be considered "typical". For partial nitrification and combined nitrification-denitrification, 3.4 g of O_2 are required for 1.0 g of NH_4-N. In this case, for a typical N loading rate of 20 kg ha^{-1} d^{-1}, 68 kg ha^{-1} d^{-1} of O_2 would be required. This is a better conversion per unit of O_2, but it still high relative to the perceived available oxygen. Less oxygen is needed when partial nitrification is combined with the ANAMMOX reaction. According to the ANAMMOX stoichiometry (Strous *et al.*, 1998; Sliekers *et al.*, 2002)

$$3NH_4^+ + 2.55O_2 + 1.7[H] \Rightarrow 1.5N_2 + 3.1H^+ + 5.3H_2O$$

1.9 g of O_2 is required for 1.0 g of NH_4-N, which includes the oxygen needed to convert ammonia to nitrite. For a typical N loading rate of 20 kg ha^{-1} d^{-1}, 39 kg ha^{-1} d^{-1} of O_2 would be required. This is much less than the oxygen requirement for standard nitrification/denitrification. However, there are other demands for oxygen in the system such as heterotrophic metabolism. While these organisms have been found in many natural environments including wastewater treatment systems and grown in sufficient quantities for scale-up reactors, it is still unknown the extent of these reactions in animal waste systems. Much of the research has been conducted under laboratory conditions with mineral media or relatively low organic content. Research is needed to better understand how the microbes and the ammonia oxidizing reactions compete in the ecology of varied wetland systems.

Riparian zones

At the watershed level, N will generally be applied to the land even after constructed wetland treatment, and agricultural systems are inherently incomplete in their assimilation of N because of variation in crop production, rainfall, temperature, and other environmental variations. Thus, some N will be discharged toward waterways. Fortunately, many natural riparian zones exist, and they are effective in buffering streams from potential N pollution (Peterjohn and Correll, 1984; Hill, 1996; Lowrance *et al.*, 2002). Their physical, chemical, and biological processes can function in assimilation and transformation of contaminants before they can be transported into stream waters. Their function has been documented throughout the world in various ecosystems to include deciduous forest in the Atlantic Coast of the USA (Peterjohn and Correll, 1984; Lowrance *et al.*, 1984; Jacobs and Gilliam, 1985; Jordan *et al.*, 1993; Hanson *et al.*, 1994; Stone, *et al.*, 1998), fens of Scandinavia (Brusch and Nilsson, 1993), grass areas of New Zealand (Cooper, 1990), forest and grass areas of Canada (Hill *et al.*, 2000), and forest and grass areas of England (Haycock and Pinay, 1993). The riparian wetland areas of the world are clearly recognised to diverge from one another, and there is often considerable variation among riparian zones of a watershed. For instance, denitrification potentials were determined in six of the major landscape structures (riparian soils of both meandering and braided streams, peat lands, coniferous flats, alder slopes, and groundwater seeps) of the Lake Nerka catchment in southwest Alaska. Here, Pinay *et al.* (2003) found significant potential denitrifying activity in all the soils of the main landscape patch types of the Lake Nerka catchment. The lowest potential denitrifying activity was measured in the peat lands. The areas with the highest denitrification were coniferous flats, groundwater seeps, and riparian soils of meandering streams. Studies by Florinsky *et al.* (2004) suggest that digital terrain models can be used to predict the spatial distribution of the microbial biomass and amount of denitrifying enzyme in the soil.

While recognizing the divergence, riparian zones with significant denitrification typically have hydrological and biological conditions that promote N removal; effective riparian zones generally contain shallow groundwater, an active plant community, massive and dynamic soil microbial populations, and hydric soils (Lowrance *et al.*, 1984; Ambus and Lowrance, 1991; Hill *et al.*, 2000; Flite *et al.*, 2001). Often the removal of N from the point of entry into the riparian zone to the point of stream entry is more than 90% (Peterjohn and Correll, 1984; Jacobs and Gilliam, 1985; Jordan *et al.*, 1993; Lowrance *et al.*, 1995). Thus, the interconnected cycling of nutrients from the groundwater into plant biomass with the subsequent soil deposit and decomposition of the leaf litter is important for both the translocation of N and microbial energy from carbon (Ambus and Lowrance, 1991).

The N cycling of riparian zones is particularly important when they are contiguous to heavily loaded animal waste application fields. Lowrance and Hubbard (2001) documented the N removal in various combinations of grass and forested wetland zones when swine wastewater was applied via overland flow. Although they found effective removal of N at high rates, they also noted the accumulation of ammonia in soil profiles along the entire length of the overland flow slope. Vellidis *et al.* (2003) also reported the importance of a restored riparian zone in the mitigation of N (> 50%) from a manure application area. Sloan *et al.* (1999) reported that stream N concentration was elevated as a stream passed by a riparian zone contiguous to a swine wastewater spray field despite high levels of denitrification in the riparian zone.

The importance of effective riparian zones was also reported for a watershed with significant livestock by Hunt *et al.* (2004). They found that the riparian zone was able to lower the N of the groundwater sufficiently as it moved from the spray field to a stream, but it was not completely effective in the removal of N. The stream contained 12 mg l^{-1} of nitrate-N. Denitrification in this riparian zone was very active. Values of DEA in the riparian zone ranged from three to 1660 µg kg^{-1} hr^{-1}. DEA values were highest next to the stream and lowest next to the spray field. Throughout the riparian zone, nitrate was generally found to be the limiting factor for denitrification. Values of DEA generally decreased with soil depth; means for the surface, middle, and bottom depths were 147, 83, and 67 µg kg^{-1} hr^{-1}, respectively. They found in a stepwise regression of log DEA, soil total N was the most highly correlated factor (r^2 = 0.64). Inclusion of water table depth, soil sample depth, and distance from the spray field gave modest improvements in the predictive capability (r^2 = 0.86). These DEA values are much higher than those commonly reported for riparian zones adjoining cropland of the South-eastern USA, but they are lower than those reported for constructed wetlands used for treatment of swine wastewater in the region.

Generally, speaking the denitrification in the riparian zone proceeds to di-nitrogen gas. However, incomplete denitrification within the riparian zone can produce nitrous oxide, which has deleterious air quality properties. Conrad (1996) reported that the production and consumption of nitrous oxide was controlled by many different types of microbes, and the controlling processes of particular environments were varied and generally poorly defined. Walker *et al.* (2002) reported that emissions of nitric oxide, nitrous oxide, and ammonia were lower in an Appalachia riparian zone that was allowed to recover from over grazing relative to a companion riparian zone that continued to be grazed by cattle (24 vs. 77 kg ha^{-1} yr^{-1} of nitrous oxide). McSwiney *et al.* (2001) reported that the consumption of nitrate and the accumulation of nitrous oxide were primarily due to differences in redox conditions along a tropical rainforest catena of Puerto Rico, which included a riparian zone. Importantly, Dhondt et. al. (2004) conducted an investigation to determine the extent of potential nitrous oxide production in three different type of riparian zones (mixed vegetation, forest, and grass) of the Molenbeek River of Belgium in an effort to assess the trade off of denitrification to improve water quality versus potential air quality degradation via nitrous oxide emissions. They concluded that observed nitrous oxide emissions in riparian zones were not a significant "pollution-swapping phenomenon."

In-stream wetlands

As previously stated, agricultural systems are inherently leaky relative to N, and this can affect stream water quality depending upon the extent of N being lost and the buffering capacity of the landscape prior discharge to the stream. For instance, in the Albemarle-Pamlico Drainage Basin of North Carolina of the USA, the highest in-stream N loads were measured in predominantly agricultural drainage areas (McMahon and Woodside, 1997). Sebilo *et al.* (2003) reported the importance of both the riparian zone and the stream shore and bottom in the cycling of N. Nitrogen budgets established for large river systems reveal that up to 60% of the nitrate exported from agricultural soils is eliminated, either when crossing riparian wetlands areas before even reaching surface waters, or within the rivers themselves through benthic denitrification.

When there is very heavy application of N to agricultural land and/or where agricultural drainage waters bypass the riparian zones, the N cycling process of in-stream wetlands can be important. The role of wetlands, both natural and man-made, in improving water quality of streams, rivers, and lakes was demonstrated in Midwestern USA (Kovacic *et al.*, 2000), and they represent a very effective way of using stream ecosystems to help improve water quality (Mitsch and Cronk, 1992). Hunt *et al.* (1999) found that re-establishing a small (3.3 ha) in-stream wetland was very effective in lowering the mass load of N from an agricultural watershed (425 ha). The total N removal of the wetland was 3 kg ha^{-1} d^{-1}, which was about 37% of the total annual inflow to the wetland. The rate of N removal was much higher during warm periods, and this was related to the effect of temperature on denitrification processes. This affect can be seen in Figure 4, which shows the change of nitrate to chloride ratio during the year.

Current practices for using in-stream cycling of N have generally focused on benthic processes. However, advances in the use of floating wetlands, immobilization of specific microbes, solar energy, and microelectronics may allow for effective and aesthetically appealing technologies for improved water quality in important steams, lakes, and estuaries that adjoin agricultural watersheds.

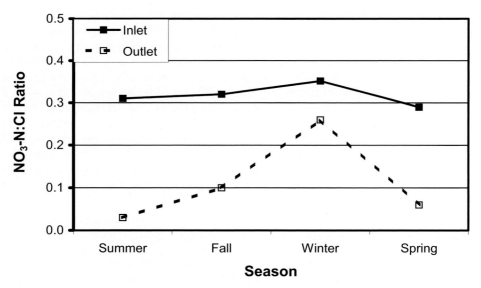

Figure 4. *Nitrate/chloride ratios for inflow and outflow of an in-stream wetland during different seasons (Hunt* et al., *1999).*

Conclusions

Understanding and effectively managing the N cycle are essential for enhanced management of N in both the constructed and natural wetlands of agricultural watersheds. The improved understanding of how aerobic and anaerobic zones and processes interacted to cycle N was exceedingly important to advancements in agricultural management and environmental protection during the later part of the twentieth century. It provided the basis for both improved management practices such as the use of constructed wetlands and riparian zones, as well as advances in the delineation of microbial processes such as ANAMMOX. Moreover, discovery of new specific aspects of the N cycle are proceeding rapidly. There is reason to be optimistic that the nexus of advancements in microbial ecology/genetics, ecological engineering, and agricultural practices will allow the N cycle of wetland systems to be understood and managed synergistically for modern agriculture.

References

Ambus P. and R.R. Lowrance, 1991. Comparison of denitrification in two riparian soils. Soil Science Society of America Journal **55** 994-997.

Bernet, N., P. Dangcong, J.-P. Delgenes, and R. Moletta, 2001. Nitrification at low oxygen concentration in biofilm reactor. Journal of Environmental Energy **127** 266-271.

Bock, E., I. Schmidt, R. Stuven, and D. Zart, 1995. Nitrogen loss caused by denitrifying Nitrosomonas cells using ammonium or hydrogen as electron donors and nitrite as electron acceptor. Archives of Microbiology **163** 16-20.

Brusch, W. and B. Nilsson, 1993. Nitrate transformations and water movement in a wetland area. Hydrobiologia **251** 103-111.

Cathcart, T.P., D.A. Hammer, and S. Triyono, 1994. Performance of a constructed wetland-vegetated strip system used for swine waste treatment. Constructed Wetlands for Animal Waste Management, edited by P. J. DuBowy and R. P. Reaves, West Lafayette: Purdue Research Foundation, pp. 9-22.

Conrad, R., 1996. Soil microorganisms as controllers of atmospheric trace gases (H_2, CO, CH_4, OCS, N_2O, and NO). Microbiological Reviews **60**(4) 609-640.

Cooper, A.B., 1990. Nitrate depletion in the riparian zone and stream channel of a small headwater catchment. Hydrobiologia **202** 13-26.

DeBusk, T.A., and J.H. Ryther, 1987. Biomass production and yield of aquatic plants. Aquatic Plants for Water Treatment and Resource Recovery, edited by K.R. Reddy and W.H. Smith, Orlando: Magnolia Publishing Inc. pp. 570-598.

Dhondt, K., P. Boechx, G. Hofman, and O. Van Cleemput, 2004. Temporal and spatial patterns of denitrification enzyme activity and nitrous oxide fluxes in three adjacent vegetated riparian buffer zones. Biology and Fertility of Soils **40** 243-251.

Flite O.P., III, R.D. Shannon, R.R. Schnabel, and R.R. Parizek, 2001. Nitrate removal in a riparian wetland of the Appalachian Valley and ridge physiographic Provence. Journal of Environmental Quality **30** 254-261.

Florinsky, I.V., S. McMahon, and D.L. Burton, 2004. Topographic control of soil microbial activity: A case study of denitrifiers. Geoderma **119** 1-2, 33-53.

Hammer, D.A. (ed.), 1989. Constructed Wetlands for Wastewater Treatment - Municipal, Industrial, and Agricultural. Lewis Publishers, Chelsea, MI.

Hanaki, K., C. Wantawin, and S. Ohgaki, 1990. Nitrification at low levels of dissolved oxygen with and without organic loading in a suspended-growth reactor. Water Resources **24** 297-302.

Hanson, G.C., P.M. Groffman, and A.J. Gold, 1994. Symptoms of nitrogen saturation in a riparian wetland. Ecological Applications **4** 750-756.

Harper, L.A., R.R. Sharpe, and T.B. Parkin, 2000. Gaseous nitrogen emissions from anaerobic swine lagoons: Ammonia, nitrous oxide, and dinitrogen gas. Journal of Environmental Quality **29** 1356-1365.

Haycock, N.E. and G. Pinay, 1993. Groundwater nitrate dynamics in grass and poplar vegetated riparian buffer strips during the winter. Journal of Environmental Quality **22** 273-278.

Hill, A.R, 1996. Nitrate removal in stream riparian zones. Journal of Environmental Quality **25** 743-755.

Hill, A.R., K.J. Devito, S. Campagnolo, and K. Sanmugadas, 2000. Subsurface denitrification in a forest riparian zone: Interactions between hydrology and supplies of nitrate and organic carbon. Biogeochemistry **51** 193-223.

Hunt P.G., A.A. Szögi, F.J. Humenik, J.M. Rice, T.A. Matheny, and K.C. Stone, 2002a. Constructed wetlands for treatment of swine wastewater from an anaerobic lagoon. Transactions of the American Society of American Engineers **45** 639-647.

Hunt, P.G., and C.R. Lee, 1975. Overland flow treatment of wastewater - A feasible approach. In: Land Application of Wastewater. USEPA, EPA 903-9-75-017, pp. 71-83.

Hunt, P.G., and C.R. Lee, 1976. Land treatment of wastewater by overland flow for improved water quality, In: Biological Control of Water Pollution, edited by J. Tourbier, and R.W. Pierson, Jr., University of Pennsylvania Press, Philadelphia, PA, pp. 151-160.

Hunt, P.G., L.C. Gilde, and N.R. Francingues, 1976. Land treatment and disposal of food processing wastewater. Land Application of Waste Materials, Soil Conservation Society of America, pp. 112-135.

Hunt, P.G., K.C. Stone, F.J. Humenik, T.A. Matheny, and M.H. Johnson, 1999. In-stream wetland mitigation of nitrogen contamination in a USA Coastal Plain stream. Journal of Environmental Quality. **28** 249-256.

Hunt, P.G., A.A. Szogi, F.J. Humenik, J.M. Rice, T.A. Matheny, and K.C. Stone. 2002. Constructed wetlands for treatment of swine wastewater from an anaerobic lagoon. Transactions of the American Society of American Engineers **45** 639-647

Hunt, P.G., T.A. Matheny, and A.A. Szogi, 2003. Denitrification in constructed wetlands used for treatment of swine wastewater. Journal of Environmental of Quality **32** 727-735.

Hunt, P.G., T.A. Matheny, and K.C. Stone, 2004. Denitrification in a coastal plain riparian zone contiguous to a heavily loaded swine wastewater spray field. Journal of Environmental of Quality **33**:2367-2374.

Hunt, P.G., M.E. Poach, G.B. Reddy, K.C. Stone, M.B. Vanotti, and F.J. Humenik, 2002b. Swine wastewater treatment in constructed wetlands. In: 10th Workshop of FAO European Cooperative Research Network on Recycling of Agricultural, Municipal and Industrial Residues in Agriculture Proc. (Slovakia), pp. 315-318.

Jacobs, T.C. and J.W. Gilliam, 1985. Riparian losses of nitrate from agricultural drainage waters. Journal of Environmental of Quality **14** 472-478.

Jetten, M.S.M., 2001. New pathways for ammonia conversion in soil and aquatic systems. Plant and Soil **230** 9-19.

Jordan, T.E., D.L. Correll, and D.E. Weller, 1993. Nutrient interception by a riparian forest receiving inputs from adjacent cropland. Journal of Environmental of Quality **22** 467-473.

Kadlec, R.H., and R.L. Knight, 1996. Treatment Wetlands. Boca Raton: Lewis Publishers, 893 pp.

Knight, R.L., V.W.E. Payne, Jr., R.E. Borer, R.A. Clarke, Jr., and J.H. Pries, 2000. Constructed wetland for livestock wastewater management. Ecological Engineering **15** 41-55.

Kovacic, D.A., M.B. David, L.E. Gentry, K.M. Starks, and R.A. Cooke, 2000. Wetlands and aquatic processes: Effectiveness of constructed wetlands in reducing nitrogen and phosphorus export from agricultural tile drainage. Journal of Environmental Quality **29** 1262-1274.

Kuenen, J.G., and L.A. Robertson, 1994. Combined nitrification-denitrification processes. FEMS Microbiology Reviews **15** 109-117.

Lee, C.R., R.E. Hoeppel, P.G. Hunt, and C.A. Carlson, 1976. Feasibility study of the functional use of vegetation to filter, dewater, and remove contaminants from dredged material. U.S. Army Engineer - Waterways Experiment Station. Technical Report D-76-4. 81 pp.

Lowrance, R.R, G.R. Vellidis, and R.K. Hubbard, 1995. Denitrification in a restored riparian forested wetland. Journal of Environmental Quality **24** 808-815.

Lowrance, R.R. and R.K. Hubbard, 2001. Denitrification from a swine lagoon overland flow treatment system at a pasture riparian zone interface. Journal of Environmental Quality **30** 617-624.

Lowrance, R.R., R.L. Todd, and L.E. Asmussen, 1984. Nutrient cycling in an agricultural watershed. II. Stream flow and artificial drainage. Journal of Environmental Quality **13** 27-32.

Lowrance, R.R., S. Dabney, and R. Schultz, 2002. Improving water and soil quality with conservation buffers. Journal of Soil and Water Conservation **57** 36-43.

Luijn, V.F., P.C.M. Boers, and L. Lijklema, 1998. Anoxic N_2 fluxes from freshwater sediments in the absence of oxidized nitrogen compounds. Water Resources **32** 407-409.

Mattingly, G.E., 1977. Experimental study of wind effects on reaeration. J. Hydraulics Div., ASCE **103**(HY3) 311.

McCaskey, T.A., S.N. Britt, T.C. Hannah, J.T. Eason, V.W.E. Payne, and D.A. Hammer, 1994. Treatment of swine lagoon effluent by constructed wetlands operated at three loading rates. In: Constructed Wetlands for Animal Waste Management, edited by P. J. DuBowy and R. P. Reaves, West Lafayette: Purdue Research Foundation. pp. 23-33.

McMahon, G. and M.D. Woodside, 1997. Journal of American Water Resources Association **33** 573-589.

McSwiney, C.P., W.H. McDowell, and M. Keller, 2001. Distribution of nitrous oxide and regulators of its production across a tropical rainforest catena in the Luquillo Experimental Forest, Puerto Rico. Biogeochemistry **56** 265-286.

Mitsch W.J., and K.K. Cronk, 1992. Creation and restoration of wetlands: Some design consideration for ecological engineering. Advances in Soil Science **17** 217-259.

Mitsch, W.J., and J.G. Gosselink, 2000. Wetlands. John Wiley & Sons, Inc., New York. 920 pp.

Mulder, A., A.A. van de Graaf, L.A. Robertson, and J.G. Kuenen, 1995. Anaerobic ammonium oxidation discovered in a denitrifying fluidized bed reactor. FEMS Microbiol. Ecol. **16** 177-184.

Muller, E.B., A.H. Stouthamer, and H.W. Verseveld, 1995. Simultaneous NH_3 oxidation and N_2 production at reduced O_2 concentrations by sewage sludge subcultured with chemolithotrophic medium. Biodegradation **6** 339-349.

O'Connor, D.J., and W.E. Dobbins, 1958. Mechanisms of reaeration in natural streams. Transactions of the American Society of American Engineers **123** 641-684.

Park, H.-D., J.M. Regan, and D.R. Noguera, 2002. Molecular analysis of ammonia-oxidizing bacterial populations in aerated-anoxic Orbal processes. Water Science Technology **46** 273-280.

Patrick W.H., Jr., and D.S. Mikkelsen, 1971. Plant nutrient behavior in flooded soil. Fertilizer Technology and Use, Soil Science Society of America, Madison, Wisconsin.

Patrick, W.H., Jr., and K.R. Reddy, 1976. Nitrification-denitrification reactions in flooded soils and sediments: Dependence on oxygen supply and ammonium diffusion. Journal of Environmental Quality **5** 469-472.

Peterjohn, W.T. and D.L. Correll, 1984. Nutrient dynamics in an agricultural watershed: Observations on the role of a riparian forest. Ecology **65** 1466-1475.

Pinay, G., T. O'Keefe, R. Edwards, and R.J. Naiman, 2003. Potential denitrification activity in the landscape of a western Alaska drainage basin. Ecosystems **6** 336-343.

Poach, M.E., P.G. Hunt, M.B. Vanotti, K.C. Stone, T.A. Matheny, M.H. Johnson, and E.J. Sadler, 2003. Improved nitrogen treatment by constructed wetlands receiving partially nitrified liquid swine manure. Ecological Engineering **20** 83-197.

Poach, M.E., P.G. Hunt, E.J. Sadler, T.A. Matheny, M.H. Johnson, K.C. Stone, and F.J. Humenik, 2002. Ammonia volatilization from constructed wetlands that treat swine wastewater. Transactions of the American Society of American Engineers **45** 619-627.

Poach. M.E., P.G. Hunt, G.B. Reddy, K.C. Stone, T.A. Matheny, M.H. Johnson, and E.J. Sadler, 2004. Ammonia volatilization from marsh-pond-marsh constructed wetlands treating swine wastewater. Journal of Environmental Quality **33** 844-851.

Poth, M., 1986. Dinitrogen production from nitrite by a Nitrosomonas isolate. Applied and Environmental Microbiology **52** 957-959.

Poth, M., and D.D. Focht, 1985. 15N kinetic analysis of N_2O production by Nitrosomonas europaea: An examination of nitrifier denitrification. Applied and Environmental Microbiology **49** 1134-1141.

Reed, S.C., 1993. Subsurface flow constructed wetlands for wastewater treatment: Technology assessment. EPA-832-R-93-001, Office of Water, USEPA, Washington, DC.

Reed, S.C., R.W. Crites, and E.J. Middlebrooks, 1995. Natural Systems for Waste Management and Treatment. 2nd ed. New York, N.Y.: McGraw-Hill.

Schmidt, I., and E. Bock, 1997. Anaerobic ammonia oxidation with nitrogen dioxide by *Nitrosomonas eutropha*. Archives of Microbiology **126** 106-111.

Sebilo, M., G. Billen, M. Grably, and A. Mariotti, 2003. Isotopic composition of nitrate-nitrogen as a marker of riparian and benthic denitrification at the scale of the whole Seine River system. Biogeochemistry **63** 35-51.

Seidel, K., 1976. Macrophytes and water purification. Biological Control of Water Pollution, edited by J. Tourbier and R.W. Pierson, Jr. University of Pennsylvania Press, Philadelphia, PA, pp. 109-121.

Sliekers, A.O., N. Derwort, J.L. Campos Gomez, M. Strous, J.G Kuenen, and M.S.M. Jetten, 2002. Completely autotrophic nitrogen removal over nitrite in one single reactor. Water Resources **36** 2475-2482.

Sloan, A.J., J.W. Gilliam, J.E. Parsons, R.L. Mikkelsen, and R.C. Riley, 1999. Groundwater nitrate depletion in a swine lagoon effluent-irrigated pasture and adjacent riparian zone. J. Soil and Water Conservation **54** 651-656.

Small, M., and C. Wurm, 1977. Data Report: Meadow/Marsh/Pond system. Brookhaven National Laboratory, Upton, New York.

Stone, K.C., P.G. Hunt, A.A. Szogi, F.J. Humenik, and J.M. Rice, 2002. Constructed wetland design and performance for swine lagoon wastewater treatment. Transactions of the American Society of American Engineers **45**(3) 723-730.

Stone, K.C., P.G. Hunt, F.J. Humenik, and M.H. Johnson, 1998. Impact of swine waste application on ground and stream water quality in an eastern Coastal Plain watershed. Transactions of the American Society of American Engineers **41** 1665-1670.

Strous, M., J.J. Heijnen, J.G. Kuenen, and M.S.M. Jetten, 1998. The sequencing batch reactor as a powerful tool for the study of slowly growing anaerobic ammonium-oxidizing microorganisms. Applied Microbiology and Biotechnology **50** 589-596.

Szögi, A.A., P.G. Hunt, and F.J. Humenik, 2000. Treatment of swine wastewater using a saturated-soil-culture soybean and flooded rice system. Transactions of the American Society of American Engineers **43** 327-335.

Tanner, C.C. and R.H. Kadlec, 2003. Oxygen flux implications of observed nitrogen removal rates in subsurface-flow treatment wetlands. Water Science and Technology **48** 191-198

USDA - Natural Resources Conservation Service. 1991. Constructed wetlands for agricultural wastewater treatment, technical requirements. Washington, DC.

USEPA, 1985. Rates, constants, and kinetics formulations in surface water quality modeling. Report EPA/600/3-85/040, Environmental Research Laboratory, Athens, GA.

Vellidis, G.R., R.R. Lowrance, P. Gay, and R.K. Hubbard, 2003. Nutrient transport in a restored riparian wetland. Journal of Environmental Quality **32** 711-726.

Walker, J.T., C.D. Geron, J.M. Vose, and W.T. Swank, 2002. Nitrogen trace gas emissions from a riparian ecosystem in southern Appalachia. Chemosphere **49** 1389-1398.

Phosphorus biogeochemistry of wetlands in agricultural watersheds

E.J. Dunne[1] and K.R. Reddy[2]
[1]Wetland Biogeochemistry Laboratory, Soil and Water Science Department, University of Florida/IFAS, 106 Newell Hall, PO Box 110510, Gainesville, FL 32611-0510, USA
[2]Soil and Water Science Department, University of Florida/IFAS, 106 Newell Hall, PO Box 110510, Gainesville, FL 32611-0510, USA

Abstract

Within agricultural watersheds, wetlands are located at the interface between terrestrial uplands and truly aquatic systems. Therefore, the processes occurring within wetland systems affect down stream water quality as water and associated nutrients such as phosphorus (P) are typically transported from upland areas to aquatic systems. This review will describe some of the common forms of P found in wetland soils/sediments and the processes responsible for P transformation and translocation. Phosphorus can enter a wetland as organic and inorganic fractions. Calcium (Ca) compounds determine the availability of inorganic P in alkaline soils while, in acidic soils iron (Fe) and aluminium (Al) controls P solubility. Inorganic P has four main fractions of decreasing bioavailability. Phosphorus sorption is one of the main processes involved in inorganic P biogeochemistry in wetland soils/sediments. Sorption is controlled by the concentration of phosphate in soil porewater and solid phases. Maximum sorption capacity of a soil can be determined using empirical models. Typically, soils only sorb P when added P in solution has a higher concentration than soil porewater. Phosphorus precipitation involves the reaction of phosphate ions with metallic cations forming solid precipitate. Inorganic P forms dominate the bioavailable fractions, whereas organic P fractions typically dominate the total P content of wetland soils/sediments. Organic P compounds can also be fractionated in decreasing order of bioavailability. At the wetland ecosystem-scale processes involved in long term P retention include: sorption on wetland substrates and the accumulation and subsequent accretion of new soil/sediment material.

Keywords: wetlands, phosphorus, adsorption, precipitation, retention, accretion.

Introduction

At the landscape-scale, streams and wetlands form a critical interface between uplands and truly aquatic systems such as lakes and rivers, as all of these systems are hydrologically linked by surface and/or subsurface flows (Mitsch and Gosselink, 1993). Since the Green Revolution of the 1960s this hydrological continuum between uplands, wetlands, and aquatic systems has become disconnected; often times, there are direct hydrological linkages between uplands and aquatic systems or hydrological inputs to wetland areas from uplands, which are extreme, in comparison to natural flows, due to changes in land drainage. Ultimately, changes in land use, for example the conversion of uplands and wetland areas to agricultural land alter natural landscape drainage patterns. In addition, modern agricultural practices typically require nutrient inputs. Phosphorus inputs can include animal feed, artificial fertilizers, animal manures and biosolids. Historically, animal manures and fertilizers were land applied to agronomic crops and grassland areas based on nitrogen (N) availability in manures and fertilizers, and plant uptake rates of receiving crops (Whalen and Chang, 2001). However, this can lead to excess applications of P with the continued application of manures and fertilizers in some areas resulting in the build up of soil P concentrations above that required for plant growth (McDowell and Sharpley, 2001). Excess applications reported for areas in the US have ranged from 1 to 9 kg P ha^{-1} (Slaton *et al.*, 2004); whereas annual P surpluses of about 20 kg P ha^{-1} yr^{-1} is reported for European farms (Edwards and Withers, 1998). Presently, many nations are adopting a more integrated approach to nutrient management whereby N and P application rates are based on crop uptake, available spread land areas, along with management practices beginning to consider nutrient fate and transport beyond the field, in an effort to become more sustainable (O'Connor *et al.*, 2005).

Both the hydrological alteration of watersheds by agriculture and the imbalance between nutrient inputs and outputs, which ultimately degrades water quality can be somewhat offset or mitigated by the use of both natural and constructed wetland ecosystems (Reddy *et al.*, 1995; Axt and Walbridge, 2003; Bruland, 2003; Koskiaho, 2003). Thus, it is important to understand the biogeochemical processes regulating P transformation and translocation within wetland ecosystems, as P is the main nutrient responsible for the eutrophication of freshwater systems.

The purpose of this paper is to provide an overview of P biogeochemistry in wetland ecosystems. It will outline (i) the forms of P in wetland ecosystems, (ii) characteristics of wetland soils and sediments (ii) inorganic P retention mechanisms, (iii) main forms of organic P in wetlands, and (iv) the processes involved in P retention by wetland systems.

Phosphorus forms in wetland ecosystems

Phosphorus retention within wetlands (either constructed or natural) can be defined as the capacity of that system to remove water column P through physical, chemical, and biological processes, and retain it in a form that is not easily released under normal environmental conditions (Reddy *et al.*, 1999; Reddy and Delaune, In press). When P enters a wetland, it is often present in both organic and inorganic forms, and the relative proportion of each depend upon soil, vegetation, geology, topography and land use characteristics of the surrounding watershed. Phosphorus forms that enter a wetland are typically grouped into: (i) dissolved inorganic P (DIP), (ii) dissolved organic P (DOP), (iii) particulate inorganic P (PIP), and (iv) particulate organic P (POP) (Figure 1).

The particulate and soluble organic P fractions may be further separated into labile and refractory components. Dissolved inorganic P is bioavailable, whereas organic and particulate forms must undergo transformation to inorganic forms before being bioavailable (Reddy *et al.*, 1995 and 1999). Both biotic (biological) and abiotic (non-biological) processes regulate P transformation and translocation. Biotic processes can include assimilation by vegetation, plankton, periphyton and microorganisms; whereas abiotic processes can include sedimentation, adsorption by soils, precipitation, and exchange processes between soil and the overlying water column.

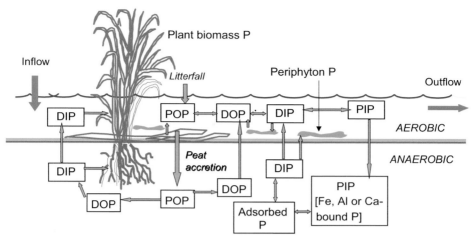

Figure 1. Schematic of phosphorus cycle in wetlands. Phosphorus fractions include dissolved inorganic P (DIP), dissolved organic P, particulate organic P, particulate inorganic P (PIP), soil adsorbed P, iron (Fe), aluminium (Al) and/or calcium (Ca) bound to PIP (Source: Reddy and DeLaune, In press).

Both organic and inorganic soluble P forms are transported within the water column, but they are transported through soil porewater. Phosphorus present in the water column is subject to changes in physicochemical characteristics of the water itself, as it flows through different reaches of the wetland and associated aquatic systems. The transport processes involved in mobilisation of P between sediment and/or soil porewater and overlying water column are advection, dispersion, diffusion, seepage, resuspension, sedimentation, and bioturbation.

Characteristics of wetland soils/sediments and overlying water columns

Oxygen diffusion in wetland soils is much slower than diffusion in a well aerated terrestrial soil such as an upland grassland soil. Slow O_2 diffusion rate in wetland water columns and high O_2 demand by soil biological communities, results in consumption of O_2 at wetland soil surfaces. This can result in the formation of a predominantly anaerobic soil that has a thin oxidised or aerobic soil layer at the soil/water interface (Mitsch and Gosselink, 1993; D'Angelo Reddy, 1994; Kadlec and Knight, 1996). Thus, two distinct soil layers are often present. They are an aerobic soil layer at the soil surface, where aerobic microbes are involved in biogeochemical reactions that use O_2 as their terminal electron acceptor, and an underlying soil layer that is predominantly anaerobic, where anaerobic microorganisms metabolise using alternative terminal electron acceptors to O_2 such as NO_3^-, Mn^{2+}, Fe^{2+}, SO_4^{2-} and CH_4^+.

In wetlands that are eutrophic such as wetlands that are used to treat runoff from agricultural areas that have high N and P levels, O_2 concentrations of the water column typically reach low levels or can become anoxic. In such instances, there is not a strong diurnal cycle of oxygen in wetland surface waters (typically during day time, high O_2 levels can occur in water column, as a result of photosynthesis by algae, while during nighttimes microbial respiration leads to a decrease of O_2 in water columns) in comparison to a natural non-impacted wetland.

Inorganic phosphorus

The availability of P in alkaline mineral wetland soils is controlled by the solubility of calcium (Ca) compounds. In mineral wetland soils that are acidic, Fe and Al minerals control the solubility of inorganic P. When soluble P is added at high concentrations to soils, insoluble mineral phosphates can form with Fe, Al and/or Ca. This process results in significant decreases in bioavailable forms of P in soils. Phosphorus loading, pH, and redox potential (a measure of electron pressure that indicates whether soils are aerobic or anaerobic) govern the stability of these phosphate minerals within soils and overlying waters.

Typically, inorganic P forms in soils are characterised based on their differential solubility in chemical extractants. Early fractionation schemes (Chang and Jackson, 1957; Petersen and Corey, 1966; Williams *et al.*, 1971) grouped soil P fractions into: (i) P present as orthophosphate ions sorbed onto surfaces (non-occluded P); (ii) P present within the matrices of P-retaining components (occluded P); and (iii) P present in discrete phosphate minerals. These schemes were modified and adapted to lake and estuarine sediments, which are typically mineral soils and later to wetland soils, which typically have high amounts of organic matter. The inorganic P fractionation schemes adopted identified the following pools: (i) exchangeable P, (ii) Fe and Al bound P, (iii) Ca and Mg bound P, and (iv) residual P (Hieltjes and Lijklema, 1980; van Eck, 1982; Psenner and Pucsko., 1988; Cooke *et al.*, 1992; Ruttenberg, 1992; Olila *et al.*, 1995; Reddy *et al.*, 1998).

Two examples in figure 2 show these different P pools that range from relatively labile fractions to more residual P fractions in selected stream sediments and wetland soils of a South Florida agricultural watershed. In this example, the inorganic P extracted with potassium chloride (KCl) represents loosely absorbed P. This fraction is considered bioavailable, therefore it can either be taken up by vegetation and/or microbes. The sodium hydroxide (NaOH) extracted P represents inorganic fractions associated with Fe and Al, and represents P that is not readily available to the overlying water column. However, under anaerobic conditions the P associated with Fe maybe released from soil to water (Wildung *et al.*, 1977; Furumai and Ohgaki, 1982; Hosomi *et al.*, 1981), whereas P bound to Al is not affected by changes in redox conditions (Moore and Reddy, 1994). The P flux from soil to water is dependent on the concentration gradient between underlying material and the overlying water column. The hydrogen chloride (HCl) fraction represents P associated with Ca and Mg and is relatively stable and not readily bioavailable. Calcium bound P such as apatite is typically found to be unavailable (Pettersson, 1986; Gunatilaka, 1988). Alkali extractable organic P includes both readily available organic P (microbial biomass P) and slowly available organic P (P associated with fulvic and humic acids). Residual P represents highly resistant organic P or unavailable mineral bound P not extracted either with a strong alkali or an acid.

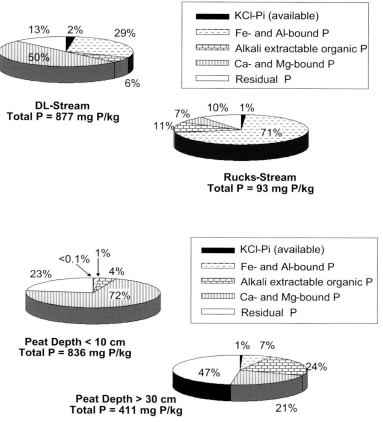

Figure 2. Phosphorus fractions in stream sediments (above) and wetland soils (below) in south Florida (Source: Reddy et al., 1995).

Soil P levels in wetlands

Total P (TP) in soils of wetlands is typically determined by either a wet ashing method or by perchloric acid digestion. Total P content of wetland soils can vary due to physicochemical characteristics, historical nutrient and hydrological loading. Values can range from 30 to 500 mg P kg^{-1} in wetlands that are not impacted by anthropogenic P loading, whereas wetlands receiving high P inputs for example from confined animal operations, can be up to 10,000 mg P kg^{-1}.

Adsorption-desorption in wetland soils and sediments

Adsorption refers to movement of soluble inorganic P from soil porewater to soil mineral surfaces, where it accumulates without penetrating the soil structure. Phosphorus adsorption capacity of a soil generally increases with clay content or mineral components of that soil (Rhue and Harris, 1999). Absorption is where soluble inorganic P penetrates into the solid phase and desorption refers to the release of adsorbed inorganic P from mineral surfaces into soil porewater. The balance between P adsorption and desorption maintains the equilibrium between solid phase and P in soil porewater. This phenomenon is defined as phosphate buffering capacity, which is analogous to pH buffering capacity of a soil (Barrow, 1983; Froelich, 1988; Rhue and Harris, 1999).

The sorption of P by soil is controlled by the concentration of phosphate in soil porewater and the ability of the solid phase to replenish phosphate into soil porewater. When soil particles become saturated with P, and soil porewater has low concentrations of P, there is a net movement of P from soil to soil porewater until there is equilibrium between soil and soil porewater P concentrations. Sorption is generally described as a two-step process:
1. Phosphate rapidly exchanges between soil porewater and soil particles or mineral surfaces (adsorption)
2. Phosphate slowly penetrates into solid phases (absorption). Similarly, desorption of P can also occur in a two-step process.

Phosphorus sorption isotherms

The phosphorus sorption capacity of a soil is typically measured using a sorption isotherm in the laboratory by mixing a known amount of soil or sediment with a solution containing a known range of P concentrations. The mixtures are equilibrated for a fixed period (usually 24 hours) at a constant temperature under continuous shaking (Nair *et al.*, 1984). Phosphorus not recovered in solution is assumed adsorbed onto solid phases. The term sorption used in the literature generally refers to both adsorption on the surface of the solid phase (or retaining component) and absorption by the solid phase (diffusion into the retaining component). Typically, P adsorption by soil increases with increasing soil porewater P concentration, until all sorption sites are occupied. At that point, adsorption theoretically reaches its maximum sorption capacity (S_{max}) as shown in figure 3. Maximum sorption capacity of a soil can be determined using a Langmuir model. Other models commonly used include in soil chemistry include Freundlich and Tempkin models (Reddy *et al.*, 1999; Rhue and Harris, 1999).

A sorption equilibrium between soil and soil porewater P concentrations can be achieved when the concentration of P in soil porewater does not change; therefore the rates of adsorption and desorption are similar, as there is no net movement of P from soil to water or vice versa. This is known as the equilibrium phosphorus concentration (EPC_o). At this point, soils exhibit their maximum buffering capacity for changes in soil porewater P concentrations (Bridgham *et al.*, 2001). At low P concentrations, the relationship between adsorption and soil porewater P concentration is linear. The intercept on the y-axis, as indicated by S_o (Figure 3) is the P adsorbed or desorbed under ambient conditions. If P is added to soil at concentrations lower than the soil porewater EPC_o, then that soil will tend to release P, until a new equilibrium is reached and vice-versa. It should be remembered that soils adsorb P only when added P concentration in solutions are higher than soil porewater EPC_o. Similarly, at the wetland-scale, if the water entering a wetland has P concentrations below EPC_o, then that soil (in theory) should tend to release P to overlying water. If the water entering a wetland has a P concentration greater than the EPC_o determined, then that soil should tend to retain P, as the P concentration gradient is from overlying water to underlying soil/sediment.

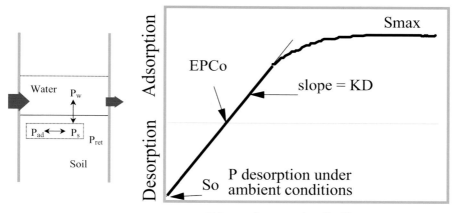

Phosphorus in Soil Porewater

Figure 3. Schematic of phosphorus sorption isotherm. The maximum sorption is known as the S_{max} and is estimated using the Langmuir sorption model. The equilibrium phosphorus concentration (EPC_0) is the solution concentration, where adsorption is equal to desorption; therefore there is no net movement of P. Phosphorus adsorbed or desorbed under ambient conditions is commonly referred to as native adsorbed P (S_0) (Adopted from Reddy and DeLaune, In press).

Chemical precipitation reactions in wetland soils and sediments

Precipitation can refer to the reaction of phosphate ions with metallic cations such as Fe, Al, Ca, or Mg, forming amorphous or poorly crystalline precipitate solids. These reactions typically occur at high concentrations of either phosphate or the metallic cations (Rhue and Harris, 1999). An example of some precipitation reactions under alkaline and acid soil conditions are shown in equations 1 and 2.

$$Ca^{2+} + HPO_4^{2-} \rightarrow CaHPO_4 \text{ [precipitate]} \tag{1}$$
$$Fe^{3+} + PO_4^{3-} \rightarrow FePO_4 \text{ [precipitate]} \tag{2}$$

In calcium dominated wetlands such as areas in the Florida Everglades, the presence of high Ca^{2+} content in water columns can result in the formation of phosphate precipitates such as: calcium phosphate (Equation 1) and hydroxyapatite. Long-term P accumulation in portions of the Florida Everglades was linearly correlated with Ca accumulation, suggesting the possibility of P and $CaCO_3$ interactions (Reddy *et al.*, 1993). Dissolution and solubilisation of solid precipitates can also take place. Dissolution occurs when the concentration of any one reactant decreases below the solubility product of that compound. Of some 200 phosphate minerals identified, only a few are of significance in wetlands. Strengite ($FePO_4$), variscite ($AlPO_4$), and vivianite ($Fe_3 (PO_4)_2$, that can occur in wetlands dominated by acidic soils, whereas beta tricalcium phosphate ($B-Ca_3(PO_4)_2$) and hydroxyaptatite ($Ca_5 (PO_4)_3 OH$) can occur in soils dominated by Ca.

Phosphorus retention and release in wetland soils and sediments

Abiotic P retention by wetland soils is regulated by various physicochemical properties present within wetland environments and they can include: pH, redox potential, Fe, Al, and Ca content of soils, organic matter content, P loading, hydraulic loading, and ambient P content of soils. Several researchers report significant correlations between amorphous and poorly crystalline forms of Fe and Al, which are typically extracted with an ammonium oxalate extraction, with P retention (sorption) by mineral soils (Berkheiser *et al.*, 1980; Khalid *et al.*, 1977; Richardson, 1985; Walbridge and Struthers, 1993; Gale *et al.*, 1994; Reddy *et al.*, 1998). It is important to note that in terrestrial soils such as those in grassland pastures, Fe and Al are typically found in crystalline forms, whereas in wetland soils, these ions typically occur in amorphous forms, which have greater surface areas for P sorption reactions to occur. Also, the presence of organic matter in wetland soils and sediments is important in P retention processes (Gale, 1994; Reddy *et al.*, 1998). Iron and Al that complex with organic matter may be responsible for such a relationship, suggesting an indirect positive effect of organic matter complexes on P retention (Syers *et al.*, 1973; Zhou *et al.*, 1997; Rhue and Harris 1999). In contrast, organic matter in terrestrial soils typically occludes P retention mechanisms with negative correlations between P sorption and organic matter reported (Dubius and Becquer, 2001).

In addition to the physicochemical characteristics of soils and sediments, the soil/sediment water interface in wetlands is important in regulating P dynamics. Processes affecting P exchange at the soil/sediment water interface include: (i) diffusion and advection due to wind-driven currents in overlying waters, (ii) diffusion and advection due to flow and bioturbation, (iii) processes within the water column itself (mineralization, sorption by particulate matter, and biotic uptake and release), (iv) diagenetic processes (mineralization, sorption and precipitation dissolution) in bottom sediments, (v) the presence of oxygen at the soil/sediment-water interface, and (vi) P flux from water column to soil mediated by evapotranspiration by vegetation.

Within overlying wetland waters and within soil porewater, the solubility of P is influenced by pH and redox potential (Eh). In a pH range of 5-8, P solubility is low at an Eh of about 300 millivolts (mV), which can result in low P concentration in soil solutions (Patrick, 1968; Patrick *et al.*, 1973). However, as the Eh decreases from 300 mV to -250 mV, P solubility increases at all pH levels, resulting in high P concentration in soil porewater (Ann *et al.*, 2000). Phosphorus solubility is highest under low pH and low Eh conditions. Under acidic conditions, an increase in P solubility is primarily due to reduction of ferric phosphate (Equation 3).

$$FePO_4 + H^+ + e^- = Fe^{2+} + HPO_4^{2-} \tag{3}$$

In soils dominated by Fe minerals, reduction of the soluble ferrous oxyhydroxide compounds results in amorphous "gel-like" reduced ferrous compounds with larger surface area than crystalline oxidised forms, as previously mentioned. A reduced soil (that has low or no availability of oxygen; therefore low redox potential) has many more sorption sites, as a result of the reduction of insoluble ferric oxyhydroxide compounds to more soluble ferrous oxyhydroxide compounds (Patrick and Khalid, 1974). Even though reduction increases sorption sites, these sites have lower P bonding energies for phosphate than do the smaller number of sites available in aerobic soils such as terrestrial grassland soils. Thus, a reduced soil will adsorb a large amount of P with a low bonding energy, thus desorption potential is high, while an oxidised soil will adsorb less P, but hold it more tightly (Patrick and Khalid, 1974). The proximity of aerobic and anaerobic interfaces in wetland soils can promote oxidation-reduction of Fe and its regulation of P solubility to the water column (Figure 4).

In sulphate dominated wetlands, production of hydrogen sulphide (H_2S) (through biological reduction of sulphate (SO_4^{2-}) and formation of ferrous sulphides may preclude P retention by ferrous iron (Caraco *et al.*, 1991). In Fe and Ca dominated systems, Moore and Reddy (1994) observed that Fe oxides likely control the behaviour of inorganic P under aerobic conditions, while Ca-phosphate mineral precipitation governs the solubility under anaerobic conditions.

During an annual timescale both constructed and natural wetland soils and sediments can undergo periods of water flooding, saturation and dry down. Studies undertaking periods of drying anaerobic soils report that phosphate buffering capacity of soils and sediments decreases upon drying (Twinch, 1987; Qui and McComb, 1994; Baldwin, 1996), while others show an increase in the degree of phosphate adsorption upon drying (Barrow and Shaw, 1980). In mineral wetland soils, drying potentially decreases the degree of hydration of iron-hydroxide gels, hence increasing surface area, resulting in increased P sorption. However, McLaughlin *et al.* (1981) observed that drying synthetic iron and aluminium oxyhydroxide increased crystallinity and decreased P sorption capacity. Under flooded and drained conditions, Sah *et al.* (1989a, b, and c) showed an increase in amorphous iron suggesting that a greater surface area was present for P sorption.

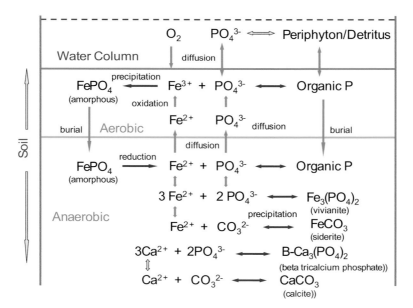

Figure 4. Phosphorus retention and release processes in water column, aerobic soil layer and anaerobic soil layer of a wetland ecosystem (Source: Moore and Reddy, 1994).

Organic phosphorus

Wetland soils are often characterised by the accumulations of organic matter (Mitsch and Gosselink, 1993; Kadlec and Knight, 1996). Due to low mineral matter content and high organic matter content, a large proportion of P in wetlands is stored in organic forms (Reddy *et al.*, 1998; Newman and Robinson, 1999). Sources of organic P in wetlands include the organic P associated with detrital organic matter from macrophytes, algae/periphyton/microbes and organic matter in incoming wetland waters. In terms of a wetland within an agricultural watershed, other organic P compounds can include organic P forms associated with herbicides, pesticides, fungicides and incoming agricultural drainage waters.

Forms of organic phosphorus

Organic P forms can be generally grouped into: (1) easily decomposable organic P (nucleic acids, phospholipids and sugar phosphates), and (2) slowly decomposable organic P (inositol phosphates or phytin). Similar to inorganic P fractions, the fractions of organic P can be conventionally distinguished by sequential and nonsequential alkali and acid extraction and fractionation procedures. They can be classed in decreasing order of bioavailability from; microbial biomass P, labile organic P, fulvic acid bound P, humic acid bound P, and residual organic P. Although these separations allow some evaluation of properties, the organic P fractions can be obscure and variable. Other methods such as bioassays with microbes (bacteria and algae) can yield insights into the bioavailability of organic P fractions. In general, large quantities of organic P can be immobilised in wetland soils, and only a small portion of the total organic P content is bioavailable. A major portion of organic P is stabilised in relatively recalcitrant organic P compounds.

Plant derived organic phosphorus

Phosphorus is a major constituent of macromolecules, particularly in nucleic acids (DNA and RNA), in phospholipids of biological membranes, and as monoesters of a variety of compounds such as those involved in biochemical pathways. In growing microorganisms, more than half of the organic P is in nucleic acids. In plants, inositol hexaphosphate can form a major storage compound for P, particularly in plant seeds. Higher aquatic wetland plants assimilate most of their P by the root-rhizome system from the wetland soil/sediment (Wetzel, 2001). Rates of P uptake and excretion by roots and leaves of submersed macrophytes, (macrophytes are plants that grow in or on the margins of water) are dependent on the P concentrations of soil porewater. Algae, bacteria, and other organisms attached to foliage can assimilate a large proportion of P that is directly removed from the water column (Moeller *et al.*, 1988). During the growing season, about 25 to 75% of the above ground P present in wetland macrophytes is translocated to rooting tissues (Granéli and Solander, 1988; Wetzel, 2001). With vegetation senescence, nutrients and residual fractions of P can leach from plant tissue. The release of P from this material is often rapid and can release from 20 to 50% of total P content in a few hours and 65 to 85% during longer periods (days). Rates of leaching are often greater from roots than from leaves. The residual detrital material is deposited on wetland soil surfaces, and becomes an integral part of the soil, thus providing long-term storage.

Organic phosphorus sorption and precipitation reactions

Organic P is readily sorbed onto clays and soil organic matter. Inositol phosphates are sorbed to clays to a greater extent than simple sugar phosphates, nucleic acids, and phospholipids. The extent and the rate of sorption depend on soil physicochemical properties and the molecular size of organic P. In acid soils, inositol phosphate is regulated by the amount of Fe and Al oxides, while under neutral and alkaline soils inositol phosphate is regulated by organic matter, clays, and Ca minerals. Phyllosillicates such as the clay mineral montmorillonite has greater affinity to sorb organic P than the other minerals such as illite, and kaolinite.

Organic P compounds can form complexes with metallic cations. The ability to form complexes depends on the number of phosphate groups in organic P compounds. For example, inositol phosphates have a greater ability to form complexes than nucleic acids and phospholipids. Similar to humic and fulvic acid complexes with metals, organic P compounds can form stable complexes with Cu^{2+}, Zn^{2+}, Ni^{2+}, Mn^{2+}, Fe^{2+} and Ca^{2+}. These complexes are stable at neutral pH. Although, stability of these complexes in wetland soils is not known, soil anaerobic conditions typically increases dissolved organic matter and associated organic P, thus increasing complexation with metallic cations. In minerals soils, reduction of ferric iron to ferrous iron can decrease potential complexation of organic P, whereas oxidation of ferrous iron to amorphous ferric oxyhydroxide can increase organic P complexation.

Phosphorus retention in wetland systems

The mechanisms that remove P in emergent vegetated wetlands both natural and constructed include: sorption on antecedent substrates, storage in biomass, and the formation and accretion of new sediments and soils (Kadlec, 1997). Thus, when evaluating a wetland ecosytem to retain P, all these components should be quantified. The first two processes are saturable, meaning they have a finite capacity and therefore cannot contribute to long-term, sustainable P removal. It is the saturation of these two pools that lead to the advance of a P front such as evident in Water Conservation 2A of the Florida Everglades. New P additions of incoming waters typically create elevated P concentrations in wetland inflow areas. A P gradient exists between inflow points and the downstream unaffected area, both in wetland surface waters and in soils (DeBusk *et al.*, 1994 and 2001). This phenomenon has been observed in both constructed and natural marsh ecosystems, for large and small additional P loadings. Phosphorus gradients are observed to lengthen, and consequently become less steep. For stable annual additions, the position of the gradient achieves a stable position.

The third and main process responsible for long term sustainable P retention is the accretion of new soils and sediments (Richardson *et al.*, 1999). This process has been verified to operative over a wide range of climatic and geographical conditions and accretion rates reported can vary from 0.4 - 4 g P m^{-2} yr^{-1} in natural wetlands (Mitsch, 1992; Faulkner and Richardson, 1989; Craft and Richardson, 1993) depending on degree of impaction. Within the USA, a long-term data base on a northern peat wetland used for wastewater treatment demonstrates that during a 30 year study period the wetland removed over 90% of the added P; peat accreted at a rate of 2-3 mm yr^{-1} during operation (Kadlec, 1993).

Conclusions

Within agricultural watersheds wetlands can be located between terrestrial uplands such as grassland and tillage areas and truly aquatic systems such rivers, streams, lakes and estuarine areas. Contaminant and nutrient loss such as phosphorus (P) from agriculture is typically transported from upland areas to aquatic systems. Therefore, due to the landscape position of wetlands, the processes occurring within wetlands, will affect down stream water quality and should be considered in watershed water quality management.

Phosphorus biogeochemistry in wetlands involves a complex of physical, chemical and biological processes occurring in wetland water columns, soils/sediments and biomass. Phosphorus that enters a wetland is in either inorganic and/or inorganic forms. Inorganic forms include exchangeable P, P bound to Ca/Mg, P bound to Fe and Al and remaining fractions of P that are residual. Phosphorus sorption and precipitation are two of the main processes in P retention by mineral wetland soils and sediments. Some of the controlling factors of sorption (which refers to both ad- and absorption) are the concentration of phosphate in soil porewater and the ability of the solid phase to replenish phosphate into porewater. Typically, soils sorb P only when added P concentration in solutions is higher than soil porewater concentrations. Precipitation involves the reaction of phosphate ions with metallic cations (both of which have to be present in high concentrations) forming precipitate solids. In general, abiotic P retention by wetland soils is regulated by various physicochemical properties such as pH, redox potential, Fe, Al, and Ca content of soils, organic matter content, P loading, and ambient P content of soils.

Organic P fractions can be the major portion of the total P content in wetland soils/sediments and usually comprises more than half of the soil total P content. Organic P fractions are also classed in decreasing order of bioavailability. Some organic P fractions can be hydrolysed to bioavailable forms such as fulvic acid bound P, whereas residual organic P is highly resistant.

The mechanisms that remove P at the emergent vegetated wetland-scale, which may be a constructed or a natural system within an agricultural watershed include: sorption on antecedent substrates, storage in biomass, and the formation and accretion of new sediments and soils. The first two processes have a finite capacity to retain P, while the third, accretion of soils and sediments is a sustainable long-term process and the rate of which, depends on nutrient impaction of the receiving wetland.

References

Ann, Y. K.R. Reddy and J.J. Delfino, 2000. Influence of chemical amendments on P retention in a constructed wetland. Ecological Engineering **14** 157-167.

Baldwin, D. S., 1996. Effects of exposure to air and subsequent drying on the phosphate sorption characteristics of sediments from a eutrophic reservoir. Limnology and Oceanography **41** 1725-1732.

Barrow, N. J., and T. C. Shaw, 1980. Effect of drying on the measurement of phosphate adsorption. Communications in Soil Science and Plant Analysis. **11** 347-353.

Barrow, N.J., 1983. On the reversibility of phosphate sorption by soils. Journal of Soil Science. **34** 751-758.

Berkheiser, V.E., J.J Street, P.S.C. Rao, and T.L. Yuan, 1980. Partitioning of inorganic orthophosphate in soil-water systems. CRC Critical Reviews in Environmental Control. 179-224.

Bridgham, S.D., C.A. Johnston, J. P. Schubauer-Berigan, and P. Weishampel, 2001. Phosphorus sorption dynamics in soils and coupling with surface and pore water in riverine wetlands. Soil Science Society of America Journal. 65:577-588.

Bruland, G. L., M.F. Hanchey, and C.J. Richardson, 2003. The effects of agriculture and wetland restoration on hydrology, soils and water quality of a Carolina bay complex. Wetlands Ecology and Management. **11** 141-156.

Caraco, N., J. Cole, and G. E. Likens, 1990. A comparison of phosphorus immobilization in sediments of freshwater and coastal marine systems. Biogeochemistry **9** 277-290.

Chang, S.C., and M.L. Jackson, 1957. Fractionation of soil phosphorus. Journal of Soil Science **84** 133-144.

Cooke, J.G., 1992. Phosphorus removal processes in a wetland after a decade of receiving a sewage effluent. Journal of Environmental Quality. **21** 733-739.

Craft, C. B. and C. J. Richardson, 1993. Peat accretion and phosphorus accumulation along a eutrophication gradient in the northern Everglades. Biogeochemistry. **22** 133-156.

D'Angelo. E.M., and K.R. Reddy, 1994. Diagenesis of organic matter in a wetland receiving hypereutrophic lake water. I. Distribution of dissolved nutrients in the soil and water column. Journal of Environmental Quality 23 925-936.

Dubus, L.G., and T. Becquer, 2001. Phosphorus sorption and desorption in oxide-rich Ferralsols of New Caledonia. Australian Journal of Soil Research **39** 403-414.

Edwards, A.C., and P.J.A. Withers, 1998. Soil phosphorus management and water quality: a UK perspective. Soil Use and Management **14** 124-130.

Faulkner, S.R., and C.J. Richardson, 1989. Physical and chemical characteristics of freshwater wetland soils, In: Constructed Wetlands for Wastewater Treatment: Municipal, Industrial and Agricultural, edited by D.A. Hammer, Lewis Publishers, Inc., Michigan, pp. 41-72.

Froelich, P.N., 1988. Kinetic control of dissolved phosphate in natural rivers and estuaries: A primer on the phosphate buffer mechanism. Limnology and Oceanography **33** 649-668.

Furumai, H., and S. Ohgaki, 1982. Fractional composition of phosphorus forms in sediments related to release. Water Science and Technology **14** 215-226.

Gale, P.M., K.R. Reddy, and D.A. Graetz, 1994. Phosphorus retention by wetland soils used for treated wastewater disposal. Journal of Environmental Quality **23** 370-377.

Granéli, W. and D. Solander, 1988. Influence of aquatic macrophytes on phosphorus cycling in lakes. Hydrobiologia **170** 245-266.

Gunatilaka, A., 1988. Estimation of the available P-pool in a large freshwater marsh. Arch. Hydrobiol. Beih. Ergebn. Limnol., **30** 15-24.

Hieltjes, H.M. and L. Lijklema, 1980. Fractionation of inorganic phosphates in calcareous sediments. Journal of Environmental Quality **9** 405-407.

Hosomi, M., M. Okada, and R. Sudo, 1981. Release of phosphorus from sediments. Verh. Internat. Verein. Limnol. **21** 628-633.

Kadlec, R. H., 1997. An autobiotic wetland phosphorus model. Ecological Engineering **8** 145-172.

Kadlec, R.H. and R.L. Knight, 1996. Treatment Wetlands. Lewis Publishers, Boca Raton, FL. 893 pp.

Khalid, R. A., W. H. Patrick, Jr., and R. D. DeLaune, 1977. Phosphorus sorption characteristics of flooded soils. Soil Science Society of America Journal **41** 305-310.

McDowell, R.W., and A. N. Sharpley, 2001. Approximating phosphorus release from soils to surface and subsurface drainage. Journal of Environmental Quality **30**:508-520.

McLaughlin, J. R., J. C. Ryden, and J. K. Syers, 1981. Sorption of inorganic phosphate by iron and aluminum containing components. Journal of Soil Science **32** 365-375.

Mitsch, W.J., 1992. Landscape design and the role of created, restored and natural riparian wetlands in controlling nonpoint source pollution. Ecological Engineering **1** 27-47.

Moeller, R. E., J. M. Burkholder, and R. G. Wetzel. 1988. Significance of sedimentary phosphorus to a submersed freshwater macrophyte (*Najas flexilis*) and its algal epiphytes. Aquatic Botany **32** 261-281.

Moore, P. A., and K. R. Reddy, 1994. Role of Eh and pH on phosphorus geochemistry in sediments of Lake Okeechobee, Florida. Journal of Environmental Quality **23** 955-964.

Nair, P.S., T.J. Logan, A.N. Sharpley, L.E. Sommers, M.A. Tabatabai, and T.L. Yuan, 1984. Interlaboratory comparision of a standardized phosphorus adsorption procedure. Journal of Environmental Quality **13** 591-595.

Newman, S. and J. S. Robinson, 1999. Forms or organic phosphorus in water, soils, and sediments. In: Phosphorus Biogeochemistry in Subtropical Ecosystems, edited by K. R. Reddy, G. A. O'Connor, and C. L. Schelske, Lewis Publishers, Boca Raton. pp. 207-223.

O'Connor, G.A., H.A. Elliot, N. T. Basta, R.K. Bastian, G. M. Pierzynski, R.C. Sims., and J.E. Smith, Jr., 2005. Sustainable land application: an overview. Journal of Environmental Quality **34**:7-17.

Olila, O.G., K.R. Reddy, and W.G. Harris, Jr., 1995. Forms and distribution of inorganic phosphorus in sediments of two shallow eutrophic lakes in Florida. Hydrobiologia **302** 147-161.

Patrick, W. H., Jr., 1968. Extractable iron and phosphorus in a submerged soil at controlled redox potentials. Trns. Intr. Congr. Soil Sci. 8th (Bucharest, Romania) **66** 605-609.

Patrick, W. H., Jr., and R. A. Khalid, 1974. Phosphate release and sorption by soils and sediments: Effect of aerobic and anaerobic conditions. Science **186** 53-55.

Patrick, W. H., Jr., S. Gotoh., and B. G. Williams, 1973. Strengite dissolution in flooded soils and sediments. Science **179** 564-565.

Petersen, G.W. and R.B. Corey, 1966. A modified Chang and Jackson procedure for routine fractionation of inorganic phosphates. Soil Science Society of America Proceedings. **30** 563-565.

Pettersson, K., 1986. The fractional composition of phosphorus in lake sediments of different characteristics. In: Sediment and Water Interactions, edited by P.G. Sly, Springer Verlag, pp. 149-155.

Psenner, R. and R. Pucsko, 1988. Phosphorus fractionation: advantages and limits of the method for the study of sediment P origins and interactions. Arch. Hydrobiol. Beih. Ergebn. Limnol. **30** 43-59.

Qui, S., and A. J. McComb, 1994. Effects of oxygen concentration on phosphorus release from reflooded air-dried wetland sediments. Australian Journal Marine and Freshwater Research **45** 1319-1328.

Reddy, K. R., and R. DeLaune, In press. Biogeochemistry of Wetlands. CRC Press, Boca Raton, Fl.

Reddy, K. R., O. A. Diaz, L. J. Scinto, and M. Agami, 1995a. Phosphorus dynamics in selected wetlands and streams of the Lake Okeechobee Basin. Ecological Engineering **5** 183-208.

Reddy, K. R., R. G. Wetzel, and R. Kadlec, 2005. Biogeochemistry of phosphorus in wetlands. In: Phosphorus: Agriculture and the Environment, edited by J. T. Sims and A. N. Sharpley, Soil Science Society of America. pp. 263-316, Madison, WI.

Reddy, K.R., G.A. O'Connor and P.M. Gale, 1998. Phosphorus sorption capacities of wetland soils and stream sediments impacted by dairy effluent. Journal of Environmental Quality **27** 438-447.

Reddy, K.R., R.D. DeLaune, W.F. DeBusk, and M.S. Koch, 1993. Long-term nutrient accumulation rates in the Everglades. Soil Science Society of America Journal **57** 1147-1155.

Rhue, R. D. and W. G. Harris, 1999. Phosphorus sorption/desorption reactions in soils and sediments. In: Phosphorus Biogeochemistry in Subtropical Ecosystems, edited by K. R. Reddy, G. A. O'Connor, and C. L. Schelske, Lewis Publishers, Boca Raton. pp. 187-206.

Richardson, C.J., 1985. Mechanisms controlling phosphorus retention capacity in freshwater wetlands. Science **228** 1424-1426.

Ruttenberg, K. C., 1992. Development of sequential extraction method for different forms of phosphorus in marine sediments. Limnology and Oceanography **37** 1460-1482.

Sah, R. N., D. S. Mikkelsen, and A. A. Hafez, 1989b. Phosphorus behavior in flooded-drained soils. II. Iron transformations and phosphorus sorption. Soil Science Society of America Journal **53** 1723-1729.

Sah, R. N., D. S. Mikkelsen, and A. A. Hafez, 1989c. Phosphorus behavior in flooded-drained soils. III. Phosphorus desorption and availability. Soil Science Society of America Journal **53** 1729-1732.

Sah, R. N., D. S. Mikkelsen, and A. A. Hafez. 1989a. Phosphorus behavior in flooded-drained soils. I. Effects on phophorus sorption. Soil Science Society of America Journal. 53:1718-1723.

Slaton, N.A., K.R. Brye, M.B. Daniels, T.C. Daniel, R.J. Normann, and D.M. Miller, 2004. Nutrient input and removal trends for agricultural soils in nine geographic regions in Arkansas. Journal of Environmental Quality **33**:1606-1615.

Syers, J.K., R.F. Harris, and D.E. Armstrong, 1973. Phosphate chemistry in lake sediments. Journal of Environmental Quality **2** 1-14.

Twinch, A. J., 1987. Phosphate exchange characteristics of wet and dried sediment samples from a hypertrophic reservoir: Implications for the measurement of sediment phosphorus status. Water Resources **21** 1225-1230.

Van Eck, G.T.M., 1982. Forms of phosphorus in particulate matter from the Hollands Diep/Haringvliet, The Netherlands. Hydrobiologia **92** 665-681.

Walbridge M.R., and J.P Struthers, 1993. Phosphorus retention in non-tidal palustrine forested wetlands of the Mid-Atlantic region. Wetlands **13** 84-94.

Walker, W. W., 1995. Design basis for Everglades stormwater treatment areas. Water Resources Bulletin. **31** 671-685.

Wetzel, R. G., 2001. Limnology: Lake and River Ecosystems, 3rd edition, Academic Press, San Diego, 1006 pp.

Whalen, J. K., C. Chang, 2001. Phosphorus accumulation in cultivated soils from long-term annual applications of cattle feedlot manure. Journal of Environmental Quality **30**:229-237.

Wildung, R.E., R.E. Schmidt, and R.C Routson, 1977. The phosphorus status of eutrophic lake sediments as related to changes in limnological conditions - Phosphorus mineral components. Journal of Environmental Quality **6** 100-104.

Williams, J.D.H., J.K. Syers, R.F. Harris, and D.E. Armstrong, 1971. Fractionation of inorganic phosphate in calcareous lake sediments. Soil Science Society of America Proceedings **35** 250-255.

Zhou, M., R. D. Rhue, and W. G. Harris, 1997. Phosphorus sorption characteristics of Bh and Bt horizons from sandy coastal plain soils. Soil Science Society of America Journal **61** 1364-1369.

Retention of soil particles and phosphorus in small constructed wetlands in agricultural watersheds

B.C. Braskerud
Norwegian Centre for Soil and Environmental Research, NO-1432 Ås, Norway

Abstract

This paper demonstrates how small constructed wetlands (CWs) of approximately 0.1 % of the watershed area can contribute to cleaner waterways. The paper sums up more than 10 years experience of investigation and construction of wetlands in Norway, where agricultural authorities sponsor constructed wetland building costs. Four investigated CWs show that the average retention in the different wetlands varied from 45-75 % for soil particles, and 21-44 % for phosphorus (P). Retention of particles and P tends to increase in relative and absolute amounts with runoff due to erosion processes in the watershed. Higher constructed wetland (CW) effectiveness than expected is probably a combined effect of (i) clay particles entering wetlands as aggregates with higher sedimentation velocity than single particles, (ii) shallow depth which gives shorter particle settling distance, and (iii) vegetation cover preventing resuspension of sediment. Due to high hydraulic loads the redox potential in the outlet water indicated aerobic conditions. As a result the redox-sensitive P in the wetland sediment is conserved as long as sufficient amount of water flows through the CW.

Keywords: aggregate, diffuse pollution, hydraulic load, redox potential, sedimentation.

Introduction

Loss of soil particles and nutrients from arable land may harm streams and lakes. Internationally, use of CWs is regarded as an efficient and often the only applicable measure for reduction of diffuse pollution in streams. Normally a large CW-surface area is required to achieve good results. However, the small Norwegian farms and the hilly landscape, makes it impossible to set aside large areas for wetlands. Consequently, small wetlands of approximately 0.1 % of the watershed area have been constructed, even though several models (Models 1 and 2) predicted little retention of particles and P due to their small dimensions.

The retention of soil particles is a key factor, since P and many other pollutants are mainly particle bound. It is generally agreed that particle sedimentation velocity, runoff and pond or wetland surface area influence retention performance. This can be expressed in two commonly used models. For fully developed turbulence, the relative retention, E (%), for a particle is (Chen, 1975; Haan *et al.*, 1994):

$$E = 100 \left[1 - \exp(-w \, AQ^{-1}) \right] \qquad \text{(Model 1)}$$

where w is the particle settling velocity (m s^{-1}), A is the CW surface area (m^2), Q is the runoff from the pond (m3 s^{-1}) and exp is the value of e (2.718...). A model similar to (1) is the first-order area model (Kadlec and Knight, 1996):

$$C_{out} = (C_{in} - C^*)exp(-k\ AQ^{-1}) + C^* \hspace{3cm} \text{(Model 2)}$$

where C_{in} and C_{out} are concentration of pollutants in inlet and outlet (mg l^{-1}). C^* is the background value (mg l^{-1}) and k is the removal rate constant (m s^{-1}). Note that the constant k is equal to w for retention of suspended soil particles, because $E = 100\ [1 - (C_{out} - C^*)(C_{in} - C^*)^{-1}]$, and that models (1) and (2) use the inverse hydraulic load (AQ^{-1}).

For both models retention is independent of depth, as stated by Hazen in 1904. Retention increases as surface area (A) and particle sedimentation velocity (w) increases, and decreases as runoff (Q) increases. Hence, doubling the pond volume by a doubling of the surface area increases the retention, while a doubling of the depth does not affect retention. In the deep pond, the extra detention time is counteracted by a greater particle travel time (Kadlec and Knight, 1996).

The objective of the paper is to show how small CWs can contribute to cleaner waterways; sedimentation of particles and retention of P. Positive and negative aspects with Models 1 and 2 as rule of thumb in CW design are discussed. The paper presents results from four wetlands, each monitored from 5 to 10 years by water proportional composite sampling in the inlets and outlets. Sedimentation traps and sedimentation plates were placed in a total of six wetlands, to study the distribution and characteristic of the sediment. In addition, the redox potential was monitored to study whether it influenced the loss of P in one CW. For more details on the sampling programme, analyses and watersheds see Braskerud 2001a, b; 2002a, b; Braskerud, 2003; Braskerud *et al.* 2005.

Design of Norwegian wetlands

The investigated CWs were located under different temperate and cold temperate climatic and agricultural regions in Southern Norway. The examined CWs contain up to four different components (Figure 1 and Table 1). Most studies were conducted in the oldest wetland, type CW-A (Table 1).

Agricultural watershed areas varied from 22 ha in CW-G to 150 ha in CW-A. Annual precipitation varied from 750 mm to 1400 mm in CWs A and G, respectively. Average annual hydraulic load (QA^{-1}) varied from 0.66 m d^{-1} (or m^3 m^{-2} d^{-1}) in CW-G2 to 3.4 m d^{-1} in CW-D. A typical Norwegian watershed includes more than 50% forest (Table 1). At present, Norwegian

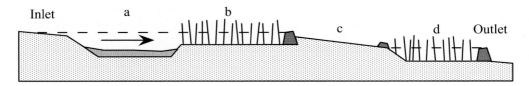

Figure 1. Components used in Norwegian constructed wetlands: (a) sedimentation pond, (b) wetland filter, (c) overflow zone covered with vegetation or stones and (d) outlet basin. Often low dams separate CW-components. Depths were originally 1 m in a, 0.5 m in b, 0 m in c and 0.5-0.8 m in d.

Table 1. Characteristics of the constructed wetlands (CWs) and farmland in the watersheds.

CW	Established	CW-size m2	% of watershed	CW- components	Arable part (%)	Production type	Clay in soil (%)
A	1990	900	0.06	a b^	17	cer./dairy	26-32
B	1990	630	0.07	a b^	11	cereals	26-29
C	1990	345	0.07	a b^	27	cereals	20-33
D	1990	265	0.03	a b^	28	cereals	18-29
F	1994	870	0.08	a^c b^b^c d^	14	meat/horse	5-7
G1	1993	460	0.21	a b^b^b^	99	dairy	5-7
G2	1993	840	0.38	G1+c^c^c^d^	99	dairy	5-7

^ is low dam or V-notch.

agricultural regulatory authorities subsidise wetland construction with approximately 70 % of the building costs. One hundred CWs were grant aided in 2002 and 85 in 2003.

Hydraulic load and soil particle retention

Retention of particles can be predicted by using Model 1. In Figure 2 the predicted retention of particles with diameter 0.6, 2, 6, 20 and 60 µm are shown as lines. The same was done by Novotny and Chesters (1981). Sedimentation velocity is estimated by Stoke's Law for water temperature (7 °C), and specific gravity of particles (2.65 g cm^{-3}). As an example, Figure 2 shows that the predicted average retention of 2 µm particles, which is the largest clay particle, should have been 17 % for the average AQ^{-1}-value 76000 m^2 m^{-3} s. The hydraulic load decreases to the right hand side in the figure. Observed retention of clay particles in composite sample events are denoted as points in Figure 2.

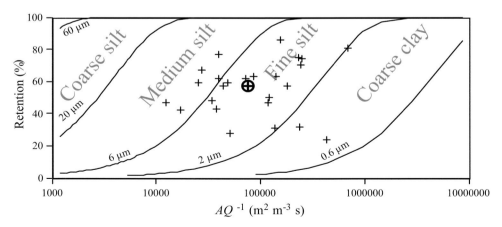

Figure 2. Predicted retention of five grain sizes (lines) and observed retention of clay (+) in CW-A as a function of inverse hydraulic load. Large symbol gives average observed AQ-1 and average observed clay retention (Modified after Braskerud, 2003).

The average observed clay retention was 57 %, which is more than three times higher than predicted by Model 1. A similar result was observed in CW-C. Since average A, Q and E is known in every composite sample, Model 1 can estimate w. The data in Figure 2 show that the clay particles behaved as fine silt and medium silt with respect to sedimentation velocity. Several investigations show that particles in streams and rivers are transported as flocs or aggregates (e.g., Droppo and Ongley, 1994). "Floc" and "aggregate" are often used synonymous, because they are difficult to distinguish (Droppo and Stone, 1994). However, their origin is not similar: aggregates are formed as a result of processes in the soil, while flocs are formed as a result of processes in running water. Suspended solids were probably dominated by aggregates (Braskerud, 2001a). Transport in streams most likely leads to a break-up of aggregates. Therefore, particle transport distance should be minimised, and this can be achieved by incorporating CWs into small watersheds.

Hydraulic load and phosphorus retention

According to Models 1 and 2, retention of all types of suspended solids decreases with increasing runoff (Q). However, as Q increased, Braskerud (2003) showed that soil particles and aggregates with higher sedimentation velocities entered the CWs. As a result, the retention often increased with increasing Q. This was also observed for total P (Figure 3). For clay particles (Figure 2) and total suspended solids, however, the positive effect of increased Q was not statistically significant. Still, the data show that retention does not decrease with increased hydraulic load as would expect. Hence Models 1 and 2 incorrectly predict retention, because they do not include the effect of soil erosion processes in the watershed.

Since CWs often have the best retention performance under storm runoff conditions, they should be located in low-order streams, even though Mitsch (1992) reported that such wetlands were somewhat unpredictable. Location beside the streams is not favourable, since by-pass-water

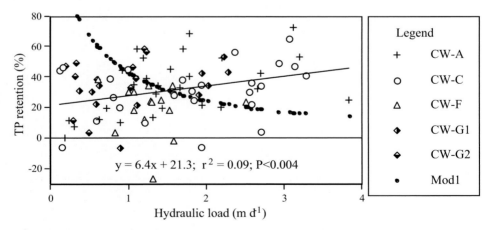

Figure 3. Observed relationship between P-retention (%) and hydraulic load (QA⁻¹) for the constructed wetlands (CWs), and according to the first-order area model, (2), (Mod1) with k = 204 and C* = 0. The data is retention per season (three months). Negative retention was net loss of phosphorus from the CWs (after Braskerud, 2002a).

will remain untreated. Even though hydraulic loads as high as 26 m d^{-1} may occur, some particle retention will occur (Braskerud *et al.*, 2000).

Table 2 summarises the results. The highest relative retention was found in the CW-G2, which had the highest surface area: watershed area ratio (Table 1), even though CW-A also had high performance. Constructed wetland-C demonstrated lowest particle retention. Smaller aggregates, due to low aggregate stability of the topsoil, are likely to explain the low retention in this CW (Braskerud, 2003). The minimum P retention was observed in CW-F, as a result of a high content of plant available P (P-AL) in the topsoil. Only 52 % of input TP was particle bound in contrast to 84 % in the other CWs. As a result, sedimentation had least effect in CW-F.

The first-order area removal constant (*k*) usually followed the same pattern as retention in individual CWs. The background value (C*) was set to 0. Since sedimentation is the most important retention process in small CWs, *k* is an estimate of the sedimentation velocity (*w*). Braskerud (2003) estimated median *w* for coarse clay particles (0.6-2 µm). It was 378 and 173 m yr-1 for CWs A and C, respectively. According to Stoke's Law *w* should have been 6-79 m yr-1. Hence, aggregates increase the sedimentation even under storm runoff events. However, aggregates with low stability have lower *w*.

Table 2. Annual retention and first-order area removal constant (k) according to Model 2 for soil particles (TSS) and phosphorus (TP) in CWs A, C, F, G1 and G2.

	Retention		*k*	Reference
	%	g m^{-2} yr^{-1}	m yr^{-1}	
TSS	45-75	16-83 x 10^3	339-727	Braskerud (2002b)
TP	21-44	26-71	124-316	Braskerud (2002a)

Phosphorus in the wetland sediment

The average TP content in the sediments was the same in CWs A and C. For wetlands F, G1, and G2 the P content in the sediments was higher (Figure 4). Generally, the P content in wetland sediments was higher than in the surrounding topsoil. In watershed F, however, the variability of the P content was higher than the topsoil in the other watersheds. In addition, the arable fields are more scattered and the P-rich topsoil may not contribute significant to the wetland sediment. As a result, it is not possible to exclude F from the general trend mentioned earlier.

Even though the P contents in the wetland sediments were high, P contents on suspended solids in the stream were even higher (median from 0.11 to 0.64 % of TSS, Braskerud, 2002a). Hence, the most P rich fraction was not retained, probably because soil particles or aggregates were too small for sedimentation.

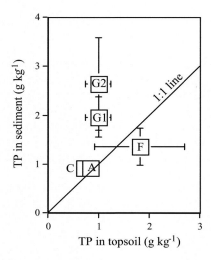

Figure 4. Relationship between total phosphorus (TP) content in the topsoil and in wetland sediments of CWs A, C, F, G1, and G2 (± one standard deviation). The 1:1 line indicates equal TP content in soil and sediment (Source: Braskerud, 2002a).

Redox potential and phosphorus retention

In general particle bound P typically settled in the small wetlands studied. The sorption behaviour of P is, however, redox-sensitive, and bound P may be remobilized in periods with low redox potential. The redox potential in the outlet water of CW-A was measured throughout a monitoring period of 3.5 years (Braskerud *et al.*, 2005). Values were always positive and often high (median 550 mV) indicating aerobic conditions. Runoff and redox potential data for 2001 are shown in Figure 5. High hydraulic loads (average 2.3 m d^{-1}) supplied the wetland with sufficient water to keep the surface water aerobic. The redox-potential in sediment without vegetation was measured occasionally with a similar electrode. It was always negative (below -200 mV) suggesting anaerobic conditions. When redox potential is below approximately 100 mV, Fe(III) is reduced to Fe(II), and iron bound P is released.

Due to the high redox potential in water, retention of P was significant, and periods with P loss from the CW, were rare. Loss was observed during less than 19 % of the total period of time. Six out of 68 episodes had a net loss of TP. The net loss was less than 5 % of the specific retention. The relative loss decreased as the redox potential increased. Penn *et al.* (2000) has shown that an oxidized micro layer exists in the upper lake sediment under well-mixed conditions in the spring and fall. This layer partly inhibits the release of sediment bound P. Our redox electrode was too large to detect the mm-thin layer. Even though the redox potential was relatively high, the oxidised micro-layer may be reduced in periods with low runoff, which may result in P release.

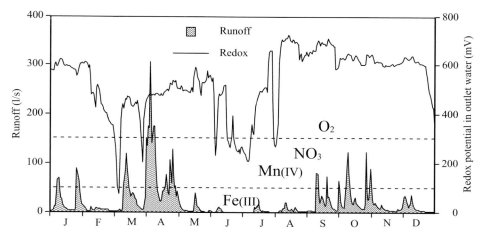

Figure 5. Daily observations of runoff and redox potential in CW-A throughout 2001. The dominating electron acceptor at a given redox potential is shown.

Wetland depth and vegetation

Phosphorus is usually attached to particles, hence sedimentation processes is very important. The positive effect of shallow wetland depths was supported in a comparative study of P retention in ponds and CWs (Uusi- Kämppä et al., 2000) (Figure 6). These ponds, located in Sweden and Finland, were often deeper than 1 meter, and vegetated only on wetland banks. The CWs were similar to those presented in this paper.

Phosphorus retention in CWs was twice the retention of ponds. Processes that may be responsible for this difference include water velocity. Water velocity on sediment surface

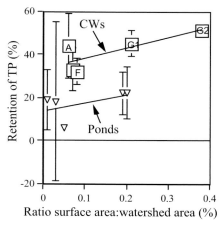

Figure 6. Retention of total phosphorus (± one standard deviation) in CWs and ponds (Modified from Uusi- Kämppä et al., 2000).

increases with a decrease in wetland water depth. Resuspension of sediment under high runoff conditions is the main reason for not building shallow CWs. Resuspension was detected in two situations in the studied CWs:

i. When CWs A, B, C and D had less than 20 % vegetation cover, approximately 40 % of the sediment was resuspended. However, as vegetation cover increased to approximately 50 %, resuspension was insignificant (Braskerud, 2001a). As a conclusion, vegetation makes it possible to utilise the positive effect of a short particle settling distance in shallow ponds, since it prevents resuspension. Thus, CW depth should be adjusted to optimal plant growth, e.g. 0.5 m or less.

ii. The overflow zones in CWs have a double function. Firstly, water is oxygenated. Secondly, soil particle retention increases due to low settling distance under low flow conditions. This should be positive for small size particles. However, as runoff increases, sediments are resuspended and lost. Concluding from this observation, an outlet basin built after overflow zones can provide a means of recapturing resuspended sediment. Overfilled CWs will probably act as overflow zones.

Surface area and hydraulic efficiency

According to Models 1 and 2, retention increases with increases in wetland surface area (A). However, the effective area involved is often less than A; if the effective volume ratio is set to 1.0 the whole volume or area is used uniformly. Figure 7 shows how the hydraulic efficiency changes for different ponds using a two-dimensional hydraulic model, MIKE-21 (Persson *et al.*, 1999). The positive effect of baffles was already known to Hazen (1904). Use of baffles is an important way to create "natural" wetlands with high hydraulic efficiency.

Surface area and maintenance

For all parameters under investigation, CW-G2 had the highest relative retention (Table 2). Hence, as the ratio of surface area (A) to watershed area increased, retention increased (Figure 6). Flooding areas is a possible way to utilise shallow depth and large A. For example, a permeable dam in the CW could be employed to flood suitable areas under storm runoff situations. Mitsch (1992) also suggested temporary flooding.

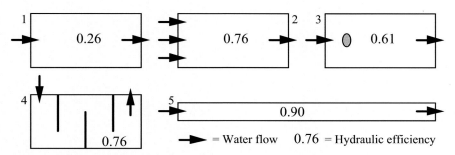

Figure 7. Hydraulic efficiency for five hypothetical ponds with a volume of approximately 2700 m3 each, and a depth of 1.5 m. The original pond (1) can be improved by spreading the inflow across the pond (2), introducing a small island in the inlet (3), using baffles (4) or by making the pond longer (5) (modified after Persson et al., 1999).

Figure 8 shows the filling rate of four CWs studied. Originally, water depth was approximately 50 cm in wetland systems. CW-D was filled in after 9 years of operation due to the accumulation of sediments. It is the smallest wetland compared to the watershed area (Table 1). The content of clay, organic matter and P are usually the same as or even larger than the original topsoil (Figure 4; Braskerud, 2002a; 2003). Often the wetland sediments reflect the topsoil of the watersheds and consist of fine, stable aggregates. Excavated sediments from filled up CWs will probably be well suited in soil mixtures and for replacement in the agricultural fields, unless it has been contaminated by industrial pollutants or agricultural diseases (Sveistrup and Braskerud, 2005).

Annual average soil loss varied from 580 to 4760 kg ha^{-1} for arable land in the F and C watersheds, respectively. Due to these excessive losses of soil from arable land the use of a sedimentation basin prior wetland treatment of incoming waters is advisable.

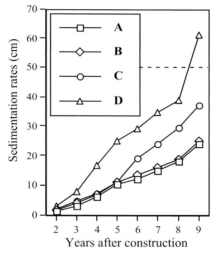

Figure 8. Cumulative sedimentation in constructed wetlands. Sedimentation was measured using sedimentation plates (Source: Braskerud, 2002b).

Conclusions

Due to the Norwegian CWs small dimensions compared to the watershed area, the hydraulic loading rate is rather high. As a result, retention models typically predict low retention of fine particles and P, which is often associated with clay particles. This paper documented that CWs are capable of retaining clay sized particles through sedimentation processes. The main factors affecting sedimentation are:
• Runoff, because erosion and transportation processes in the watershed are able to deliver large sized particles to the CWs
• Aggregates, because the clay particle settling velocity is improved. This influences retention of pollutants attached to clays.

In conclusion, CW design needs to incorporate several components for successful soil particle and P retention. Important factors include:

1. Shallow depth, which stimulates sedimentation and improves plant growth.
2. Vegetation, which increases hydraulic efficiency under storm flows and mitigates resuspension of sediments.
3. Baffles as well as input and output structures, which increases hydraulic efficiencies within CWs
4. Hydraulic loading rate
5. Constructed wetland surface area, as retention increases with increases in wetland surface area.

Acknowledgements

I thank the Norwegian Ministry of Agriculture, the Norwegian Agricultural Authority and the Research Council of Norway for their financial support. I would also like to thank H. French for her constructive criticism of the manuscript.

References

Braskerud, B.C., 2001a. The influence of vegetation on sedimentation and resuspension of soil particles in small constructed wetlands. Journal of Environmental Quality **30**(4) 1447-1457.

Braskerud, B.C., 2001b. Sedimentation in small constructed wetlands. Retention of particles, phosphorus and nitrogen in streams from arable watersheds. Dr.Scient theses 2001:10, Agric. Univ. of Norway (now: The Norwegian University of Life Sciences), Ås, Norway.

Braskerud, B.C., 2002a. Factors affecting phosphorus retention in small constructed wetlands treating agricultural non-point source pollution. Ecological Engineering **19**(1) 41-61.

Braskerud, B.C., 2002b. Design considerations for increased sedimentation in small wetlands treating agricultural runoff. Water Science and Technology **45**(9) 77-85.

Braskerud, B.C., 2003. Clay particle retention in small constructed wetlands. Water Research **37**(16) 3793-3802.

Braskerud, B.C., H. Lundekvam, and T. Krogstad, 2000. The impact of hydraulic load and aggregation on sedimentation of soil particles in small constructed wetlands. Journal of Environmental Quality **29**(6) 2013-2020.

Braskerud, B.C., T. Hartnik, and Ø. Løvstad, 2005. The effect of the redox-potential on the retention of phosphorus in a small constructed wetland. Water Science and Technology, **51**(3-4), in press.

Chen, C., 1975. Design of sediment retention basins. In: Proc. Nat. Symp. on Urban Hydrl. and Sedim. Control, University of Kentucky, Lexington, July 28-31., 285-298.

Droppo, I.G., and E.D. Ongley, 1994, Flocculation of suspended sediment in rivers of southeastern Canada. Water Resources **28** 1799-1809.

Droppo, I.G., and M. Stone, 1994. In-channel surficial fine-grained sediment lamiae. part I: physical characteristics and formational processes. Hydrological Processes **8** 101-111.

Hazen, A., 1904. On sedimentation. Transactions of the American Soc. of Civil Engineers **53** 43-71.

Haan, C.T., B.J. Barfield, and J.C. Hayes, 1994. Design hydrology and sedimentology for small catchments. Academic Press, New York.

Kadlec, R.H. and R.L. Knight, 1996. Treatment wetlands. Lewis Publishers, New York.

Mitsch, W.J., 1992. Landscape design and role of created, restored, and natural riparian wetlands in controlling nonpoint source pollution. Ecological Engineering **1**(1) 27-47.

Novotny, V. and G. Chesters, 1981. Handbook of nonpoint pollution. Van Nostrand Reinhold Company, New York.

Persson, J., N.L.G. Somes, and T.H.F. Wong, 1999. Hydraulic efficiency of constructed wetlands and ponds. Water Science and Technology **40** (3) 291-300.

Sveistrup, T. and B.C. Braskerud, 2005. Characterisation of aggregates in catchment soils and wetland sediments (in Norwegian). Grønn kunnskap, Planteforsk, Aas, Norway, **9**(2) 119-124.

Uusi-Kämppä, J., B. Braskerud, H. Jansson, N. Syversen, and R. Uusitalo, 2000. Buffer zones and constructed wetlands as filters for agricultural phosphorus. Journal of Environmental Quality **29** (1) 151-158.

Constructed wetlands to remove nitrate

R.H. Kadlec
Wetland Management Services, 6995 Westbourne Drive, Chelsea, MI 48118, USA

Abstract

Free water surface treatment wetlands are finding increasing usage for nitrate reduction in waters. An extensive database now exists, in fragmented form, in many publications and operating reports. First order models are appropriate, if implemented with due regard to hydraulic efficiency and flow patterns. Annual average rate constants are distributed around 0.25 d^{-1}. Performance is better at higher water temperatures, with a modified Arrhenius temperature factor of 1.088. Values of the tanks-in-series parameter are centered around N = 4.5. Wetland performance for nitrate removal also increases with prevention of short-circuiting and better areal contacting. Higher nitrate removal efficiencies are associated with submergent and emergent soft tissue vegetation, and lower efficiencies with unvegetated open water and with forested wetlands. Carbon availability limits denitrification at high nitrate loadings; however, wetlands produce carbon in sufficient quantities to support typical municipal and agricultural nitrate loads. Significant ancillary benefits of ecological diversity and wildlife habitat are certain to accompany the project. Economic issues include land cost and water transport.

Keywords: nitrate, wetlands, removal, model, design.

Introduction

Nitrate contributes to eutrophication of surface water. It is especially detrimental to marine aquatic systems, in which it can collapse the entire ecosystem (Diaz and Solow, 1999). Oxidised nitrogen is also important in water quality control because it is potentially toxic to infants (may result in a potentially fatal condition known as methylglobanemia) when present in drinking waters. The current regulatory criterion for nitrate in groundwater and drinking water supplies in the USA is 10 mg l^{-1}.

Both point and non-point sources contribute to the nitrogen content of waters within a typical drainage basin. In the Mississippi basin in the USA, about 60% of the water-borne total nitrogen is in the form of nitrate (Goolsby and Battaglin, 2000). The source of nitrogen is about two-thirds from agriculture, and one-third from other sources, including urban runoff, atmospheric deposition, and point sources. Nitrogen fertilisers may be applied as ammonia, of which, some fraction is unavoidably lost to infiltration. Passage through the aerobic upper soil horizon causes nitrification of ammonia to nitrate, which then exits the field via runoff or tile drainage. Similarly, pastureland typically displays nitrate runoff and drainage. Increasingly, wastewater treatment plants employ nitrification, and sometimes partial denitrification. For example, 225 wastewater treatment plants in the Illinois River basin, out of 278 total have partial or complete nitrification (State of Illinois, 2001). Their effluents are ammonia- and carbon-poor, and nitrate-rich.

The most highly oxidized forms of nitrogen in water are nitrate (NO_3-N) and nitrite (NO_2-N). The combination is commonly called oxidized nitrogen (NO_x-N), or total oxidized nitrogen (TON). Because nitrite is normally present in very low concentrations, the combination is also referred to as nitrate nitrogen, and that is the terminology adopted here. Nitrogen undergoes significant re-speciation in the environment, therefore three forms of N are important: oxidized, ammonia and organic nitrogen. The three forms are measured via laboratory analyses for total Kjehldahl nitrogen (TKN), ammonia nitrogen (NH_4-N) and nitrate (NO_x-N). Organic N is determined as the difference between TKN and NH_4-N, and total N (TN) is the sum of TKN and NO_x-N. Ammonia and organic N may be in dissolved form, or associated with suspended particulate matter. Nitrate, however, is virtually exclusively found in dissolved form. Nitrate is formed in aerobic wetland environments by nitrification of ammonia. In this paper, attention is restricted to ammonia-poor systems, and focused entirely upon the consumption and removal of nitrate within wetlands.

Nitrate is chemically stable, and persists unchanged unless utilised in microbiological nitrogen transformation processes. The microbes involved, including common facultative bacterial groups such as *Bacillus*, *Enterobacter*, *Micrococcus*, *Pseudomonas*, and *Spirillum*, are preferentially sited on submerged solid surfaces. Purely aquatic systems are therefore less efficient than wetlands, which have far larger amounts of immersed area, sometimes reaching 10 m^2 m^{-2} of wetland area (USEPA, 1999).

Nitrate uptake by wetland plants is presumed to be less favoured than ammonium uptake. Aquatic macrophytes utilise enzymes (nitrate reductase and nitrite reductase) to convert oxidized nitrogen to useable forms. The production of these enzymes decreases when ammonium nitrogen is present (Melzer and Exler, 1982). Within the wetland, nitrate is lost on its way into the sediments and root zone, due to denitrification. Thus, there may be insufficient nitrogen reaching the plant root zone. This phenomenon has been suggested as the cause of vegetation stress and death at the Trés Rios, Arizona treatment wetlands. Treatment wetlands have been proven effective in removal of nitrate (Kadlec and Knight, 1996). It is the purpose here to summarize existing performance data, interpret it into rational design criteria, and describe appropriate implementation strategies.

Nitrate removal chemistry

The chemistry of nitrate removal in conventional wastewater treatment processes has been studied in detail (U.S. EPA, 1993). An electron donor (i.e., carbon substrate) is required, which can be wastewater carbonaceous matter (carbonaceous biochemical oxygen demand or CBOD), or degradable solid or soluble organic material. Microbes essentially burn the organic matter with nitrate, but some nitrate may also be used by these bacteria for cell synthesis. The overall denitrification reaction, based on methanol (CH_3OH) as a carbon source, is summarised by the following (U.S. EPA, 1993):

$$NO_3^- + 1.08CH_3OH + 0.24H_2CO_3 = 0.056C_5H_7NO_2 + 0.47N_2 + 1.68H_2O + HCO_3^- \tag{1}$$

From the stoichiometry of equation 1, 1.07 grams of carbon are required to support the denitrification of one gram of nitrate nitrogen (Ingersoll and Baker, 1998). In the absence of this carbon source, denitrification is inhibited. Theoretically, denitrification does not occur in the presence of dissolved oxygen. However, denitrification has been observed to occur in wetland treatment systems that have measurable dissolved oxygen concentrations (Phipps and Crumpton, 1994; van Oostrom, 1994). This observation is explained by the presence of microscopic, anoxic zones that are likely to occur in biofilms and anaerobic soil layers (Reddy and Patrick, 1984).

The denitrifier populations that conduct these microbial conversions can adapt to local conditions within the wetland. Denitrifying bacteria are more abundant in higher concentrations of carbon and nitrate. At the Houghton Lake wetland (unpublished data), there was a measured increase of about a factor of 100 in the number of denitrifiers in the wastewater discharge area after the discharge commenced (10^4 increased to 10^6 per gram for litter; 10^4 increased to 10^7 per gram for soil). At Abbu Attwa, Egypt denitrifiers decreased from 10^7 to 10^5 cm^{-2} of media over the travel length of 100 m (May *et al.*, 1990). The adaptation of the sediment denitrification rate constant to higher values at higher nitrate loadings has been documented for sewage-impacted freshwater estuaries (King and Nedwell, 1987). Accordingly, rates of denitrification should not be considered constant over the spatial extent of a wetland.

Wetland hydrology and efficiency

The performance of wetlands depends in general upon the efficiency with which water contacts wetland components. Both dead zones and short-circuits impair performance, the former by reducing the effective area of the wetland, and the later by causing short contact times with substrates. The volume effects may be empirically lumped, and a volumetric efficiency (e_V) defined as:

$$e_V = \frac{V_{active}}{(LWh)_{nominal}} = \frac{\varepsilon\eta h}{h_{nominal}} \qquad (2)$$

where:
e_V = wetland volumetric efficiency, -
h = wetland water depth, m
$LWh_{nominal}$ = nominal wetland volume, m^3
V_{active} = volume of wetland containing water in active flow, m^3
ε = fraction of active volume occupied by water, -
η = fraction of area in active flow = A_{active}/A, -
τ = detention time, d

In terms of detention time,

$$\tau n = \frac{V_{nominal}}{Q} = \frac{(LWh)_{nominal})}{Q} \qquad (3)$$

It is then clear that:

$$\tau_n = e_V \bullet \tau \qquad (4)$$

Volumetric efficiency reflects ineffective volume within a wetland, compared to presumed nominal conditions. Portions of the nominal volume are blocked by submerged biomass (ε), bypassed (η), or do not exist because of poor bathymetry ($h/h_{nominal}$). Values of volumetric efficiency are generally in the range $0.7 < e_V < 0.95$, and serve to reduce actual detention times by this fraction.

Apart from these inefficiencies, there is also a distribution of detention times (DTD) in a wetland that is not typically represented by plug flow conditions (Kadlec, 1994; Kadlec and Knight, 1996). The spectrum of detention times may typically be fit with a gamma distribution, which corresponds to a tanks-in-series (TIS) concept. The number of tanks in series (N) that parameterizes a distribution is often in the range $1 < N < 5$ for single wetland cells with no internal berms. Higher N-values favor better performance, with $N = \infty$ being the plug flow, and the best performance.

Models and performance data

Design model

Data strongly suggests a first-order relation for nitrate removal, including several mesocosm and field investigations (Kadlec, 1988; Gale *et al.*, 1993; Crumpton *et al.*, 1993; Phipps and Crumpton, 1994; Spieles and Mitsch, 2000). Figure 1 shows typical field enclosure data, which follows the exponential decline expected for batch systems.

In a flow-through wetland, this first-order behavior must be combined with the DTD to produce the expected concentration profiles and exit concentrations. For a TIS distribution, nitrate removal data are then represented by:

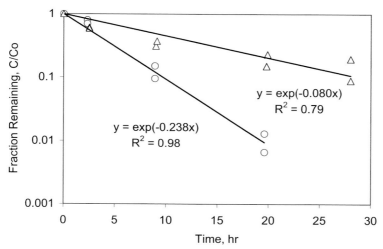

Figure 1. Reduction of nitrate in insitu field mesocosms in the Houghton lake treatment wetland. The seven-liter containers were dosed to provide starting concentrations of 90 mg l^{-1}. These were located in a cattail control area (triangles) and in the discharge area (circles).

$$\frac{C_o}{C_i} = (1 + \frac{k_v \tau_n}{N})^{-N} \qquad (5)$$

where
C_i = inlet concentration, g m^{-3}
C_o = outlet concentration, g m^{-3}
k_v = first order volumetric uptake rate constant, 1 d^{-1}
N = hydraulic efficiency parameter
n = nominal detention time, days.

Temperature effects upon denitrification are traditionally described by a modified Arrhenius temperature relation (Metcalf and Eddy, 1991):

$$k_v = k_{V20} \theta^{(T-20)} \qquad (6)$$

where
k_{V20} = first order uptake rate constant at 20°C, m yr^{-1}
T = water temperature, °C
θ = temperature factor, -

Equations (5) and (6) form the design model. Three parameters are required: k_{V20}, θ and N. Note that the nominal detention time (τ_n) has been used, which implies that the wetland volumetric efficiency has been absorbed in the k_{V20} value, and will contribute to the scatter of intersystem values.

Performance data

Information from 66 free water surface wetlands was collected from a variety of sources, including 17 from the North American Database (CH2M Hill, 1998) and the remaining 49 from published literature and project operating records. Monitoring periods ranged from one to eight years.

Thirty of the systems have been tracer tested to determine values of N, with mean ± standard error = 4.5 ± 0.4. This mean value was presumed for the remaining 36 wetlands in the determination of annual average k_v values. The median of the k_{V20} values (one point per system, regardless of the number of years) was 0.25 d^{-1}, with a standard error of 0.03. A smaller set of wetlands (33) possessed enough information to derive θ-values, which averaged 1.088 ± 0.010 (mean ± standard error).

Data may be compared to the model with these parameter values (Figure 2). Some of the scatter is due to the effect of average annual temperature, which ranged from 15 to 25 °C. However, a good portion of scatter is also due to site-specific factors, including volumetric efficiency and vegetation type.

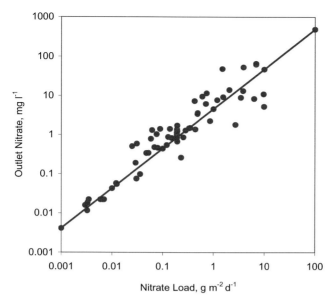

Figure 2. The relationship between nitrate load and the outlet concentration. The line is calculated from the NTIS model, with N = 4.5 and k = 0.25 d⁻¹.

Design considerations

The loading and detention required to achieve a specific goal serve to set the wetland area needed. Equation (5) is solved for the detention time needed to achieve that goal.

Load reduction versus concentration reduction

A key feature of treatment wetlands is the ability to manage the system for either concentration reduction or for mass removal, but only one at the expense of the other (Trepel and Palmeri, 2002). Wetland performance as documented above follows the rule of mass action: the removal rate of nitrate is greater at higher nitrate concentrations in the water. As a result, removal rates decrease as water passes through the treatment wetland, and nitrate concentrations are reduced. However, the actual mass of nitrate-nitrogen that is removed increases with decreasing detention time. Thus, decreasing detention time results in more tons removed, but at the expense of a lesser concentration reduction. This trade off between removal efficiency and load reduction is a key feature of wetland design for nutrient control, and may be quantified via the first order model:

$$\text{eff} = \frac{C_i - C_0}{C_i} = 1 - (1 + \frac{k\tau_n}{N})^{-N} \tag{7}$$

$$LR = LI \cdot \text{eff} = qCi \bullet (1 - (1 + \frac{k\tau_n}{N})^{-N}) \tag{8}$$

where
eff = fractional concentration reduction
LI = incoming nitrate nitrogen load, $gmN\ m^{-2}\ yr^{-1}$
LR = nitrate nitrogen load reduction, $gmN\ m^{-2}\ yr^{-1}$
q = hydraulic loading rate, $m\ yr^{-1}$.

The maximum number of tons of nitrate that can be removed in a given wetland results from a hydraulic load so high that little or no concentration reduction is achieved.

Wetland water temperatures

The meteorological driving forces that control water temperature undergo a strong seasonal cycle, and thus also the resultant wetland water temperature. Data from free water surface wetlands are therefore closely related to air temperatures. Treatment wetland information thus allows prediction of wetland water effluent temperatures to within about 1-2°C, based upon air temperatures and the times of freeze-up and thaw. To a first approximation, the water temperature during the unfrozen season is the same as the air temperature. In arid climates, evaporative cooling causes lower temperatures by a few degrees during the summer; in a wet climate, the water is a few degrees warmer than the air in mid-summer. Under-ice water temperature is 1-2 °C.

Vegetation

Vegetation serves several functions in nitrate reduction, including carbon supply and microbial attachment sites. Wetlands may contain emergent or submergent vegetation, and areas of unvegetated open water. Plants may be woody or soft-tissued. Community specificity for denitrification is expected, roughly correlated with carbon availability and the amount of immersed surface area.

The amount of total carbon in dead and decomposing biomass is on the order of 40% of the dry biomass (Baker, 1998; Ingersoll and Baker, 1998; Hume *et al.*, 2002). Not all of the total carbon produced is available for denitrifiers. Baker (1998) has suggested that the C:N loading ratio be at least 5:1 so that carbon does not become limiting, which in his work translated to 20% availability. Hume *et al.* (2002) suggest 8% availability. Presuming a carbon content of 40%, the required productivities are at the lower end of the range for emergent marshes (Kadlec and Knight, 1996). However, realisation of higher nitrate removal rates, corresponding to higher inlet concentrations, may stress the ability of the wetland to generate the required carbon energy source.

Unvegetated open water areas do not promote denitrification, resulting in k_v values about one-third those for vegetated systems (Arheimer and Wittgren, 1994). Smith *et al.* (2000) have shown nitrate removal proportional to number of shoots in a *Schoenoplectus* spp. wetland. Wetlands with woody species - shrubs and trees - also have relatively low rates of denitrification. (DeLaune *et al.*, 1996; Westermann and Ahring, 1987). Carbon limitation is the likely cause.

Either emergent or submergent vegetation can harbour epiphytic microbial biofilms on living and dead plant material (Ericksson and Weisner, 1997). However, living underwater plants produce oxygen, which inhibits denitrification. Field data do not provide clear guidance on the choice between emergent and submergent plants. Weisner *et al.* (1997) found *Potamageton* to be more effective than *Glyceria*, and *Phragmites* stands to be better than open water. Ericksson and Weisner (1997) measured very high rates of denitrification in a reservoir with dense *Potamageton pectinatus*. Conversely, Gumbricht (1993a) found low rates for *Elodea canadensis*. Toet (2002) found that emergent stands of *Typha* and *Phragmites* yielded nitrate removal rates of 98 and 287 kg ha^{-1} yr^{-1} respectively, while mixed submerged aquatics (*Elodea, Potamogeton* and *Ceratophyllum*) removed only 16 - 20 kg ha^{-1} yr^{-1}. These considerations lead to the conclusion that fully vegetated marshes with either emergent or submergent communities are the preferred option for nitrate removal. Weisner *et al.* (1994) reached this conclusion, and suggested that an alternating banded pattern perpendicular to flow would additionally provide hydraulic benefits.

Layout and siting

Efficient hydraulics requires cells in series, with intermediate collection and redistribution of water. Point inlets and outlets are to be avoided, because these cause dead zones in corners. In larger wetlands, it is desirable to provide for parallel flow paths, to allow for maintenance that may need the basin to be out of service.

Water depths for emergent plants are typically in the range of 15-45 cm. Shallower water is conducive to channelisation (short-circuiting); while deeper water may exceed the tolerance of emergent macrophytes. In event-driven wetlands, it may be necessary to provide for maintenance water in times of low runoff. This may be accomplished by positioning the wetland vertically to access groundwater, or by irrigation of the wetland in dry seasons. Emergent plants often have good tolerance for periods of dry-out, but submergent plants do not.

Seasonality

Point sources often provide a relatively steady water flow, but may possess considerable seasonality in their nitrate content. Conventional treatment plants rely on microbial nitrification, which is temperature sensitive. As a consequence, the amount of nitrate reaching a treatment wetland from such a source will vary seasonally, and be subject to seasonally variable treatment in the wetland (Figure 3).

Wetlands treating runoff experience pulse driven flows, as well as variable nitrate content. There may be no inflow at all during the dry season, and extreme flows occurring as the result of seasonally high rainfall, perhaps coupled with snow melt in cold climates. Further, no year is like any other, leading to inter-annual variability. Nitrate removal proceeds at slower rates in cold temperatures, thus a fixed percentage removal requires a lower hydraulic loading (higher detention time) under cold conditions. For example, nitrate loads appear in the rivers of the upper Mississippi basin from March through June, and again from September through November, depending upon spring thaw and fall rain timing, coupled with fertiliser application

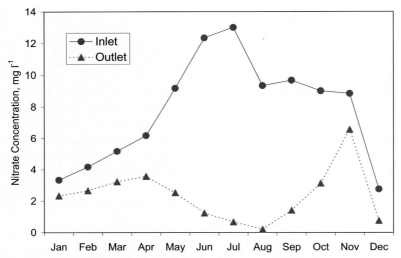

Figure 3. Nitrate removal in the Linköping, Sweden pilot wetlands (data courtesy K. Sundblad-Tonderski). The wastewater treatment plant provides more nitrate in summer, due to effective nitrification. The wetland removes more nitrate in summer, due to effective denitrification.

practices (Kovacic *et al.* 2001; Phipps and Crumpton, 1994). Temperatures at those times are typically 12-15°C. An example from the milder climate of New Zealand is shown in Figure 4.

Ancillary considerations

Treatment wetlands, by their very nature, invariably provide wildlife habitat (Knight, 1997; Worrall *et al.*, 1997; Knight *et al.*, 2001). However, habitat design goals can, and do, conflict

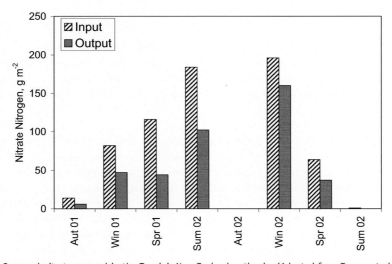

Figure 4. Seasonal nitrate removal in the Toenipi, New Zealand wetlands. (Adapted from Tanner et al., *2005).*

with water treatment goals in many cases. For example, waterfowl wetlands typically are designed with about 50% open water (Weller, 1978), but open water areas (as outlined above) impact carbon supply, which limit denitrification processes. Other potential conflicts include herbivory by rodents and fish. Beavers (*Castor canadensis*), muskrats (*Ondatra zibethicus*) and nutria (*Myocastor coypus*) have all caused extensive damage to vegetation, and damage containment dikes (Knowlton *et al.*, 2002). Bottom-foraging fish such as carp (*Cyprinus carpio*) can enter a treatment wetland if connected to a regional watercourse. Their presence can create disturbed sediments and turbid waters, which in turn impairs nitrate removal processes. Human use, such as waterfowl hunting, may not interfere significantly with nitrate removal operations, simply because of timing. Cold fall temperatures are not optimal for treatment. Passive use, for bird watching, nature interpretation or hiking is normally non-intrusive.

Discussion and conclusions

Treatment wetlands are an attractive option for controlling nitrate pollution. Both nitrified secondary effluents and stormwater runoff can be improved in water quality. Wetlands constructed for the purpose of nitrate removal are a growing application for wetland technology, which already possesses over two decades of growth and acceptance. A large experience base is now available that has been used to calibrate first generation models, and those calibrations have to a large extent converged. It is therefore possible to forecast nitrate reduction within a rather narrow uncertainty band.

The economics of implementation of treatment wetlands for nitrate control are attractive (Hey *et al.*, 2004). In the context of point source control, it is because of the large cost of conducting denitrification in a conventional treatment plant. For non-point pollution control, there is the additional factor of technological difficulty. However, regulation and permitting must be flexible enough to deal with small-scale systems without imposing unreasonable monitoring costs.

Treatment wetlands are a land-intensive option for pollution control. Sufficient land is not likely to be in public ownership, thereby requiring either purchase or lease arrangements. Large parcels are not likely to be situated near large municipalities, which raise issues of wastewater transport. An option for pollution credit trading may be necessary in such cases. In many situations, the required land may currently be in agriculture. Taking land out of production implies the need for compensation, either as a purchase, or as an income stream that replaces crop or animal revenues.

References

Arheimer, B. and H.B. Wittgren, 1994. Modelling the effects of wetlands on regional nitrogen transport. Ambio **23**(6) 378-386.

CH2M Hill, 1998. North American Treatment Wetland Database v2.0. Compact Disk, CH2M Hill, Gainesville, FL, USA.

Crumpton, W.G., T.M. Isenhart and S.W. Fisher, 1993. The fate of non-point source nitrate loads in freshwater wetlands: results from experimental wetland mesocosms. In: Constructed Wetlands for Water Quality Improvement, edited by G.A. Moshiri, Lewis Publishers, Boca Raton, FL, pp. 283-292.

DeLaune, R.D., R.R. Boar, C.W. Lindau and B.A. Kleiss, 1996. Denitrification in bottomland hardwood wetland soils of the cache river. Wetlands **16**(3) 309-320.

Diaz, R.J. and A. Solow, 1999. Ecological and Economic Consequences of Hypoxia. Topic 2 Report for the Integrated Assessment of Hypoxia in the Gulf of Mexico. NOAA Coastal Ocean Program, Washington, DC.

Eriksson, P.G. and S.E.B. Weisner, 1997. Nitrogen removal in a wastewater reservoir: the importance of denitrification by epiphytic biofilms on submersed vegetation. Journal of Environmental Quality **26**(3) 905-910.

Gale, P.M., K.R. Reddy and D.A. Graetz, 1993. Nitrogen removal from reclaimed water applied to constructed and natural wetland microcosms. Water Environment Research **65**(2) 162-168.

Gumbricht, T., 1993. Nutrient removal capacity in submersed macrophyte pond systems in a temperate climate. Ecological Engineering **2**(1) 49-62.

Hey, D.L., J.A. Kostel, A. Hurter, and R.H. Kadlec, 2004. Comparative Economic Analysis of Nutrient Removal Technologies. Final Report WERF Project 03-WSM-6CO, Water Environment Research Foundation, Alexandria, VA, 2004, 125 pp.

Ingersoll, T.L. and L.A. Baker, 1998. Nitrate removal in wetland microcosms. Water Research **32**(3) 677-684.

Kadlec, R.H., 1988. Denitrification in wetland treatment systems. In; Proceedings of the 61st Annual Conference of the Water Pollution Control Federation, Dallas, TX, 1988, Session 26, WPCF, Alexandria, VA.

Kadlec, R.H., 1994. Detention and Mixing in Free Water Wetlands. Ecological Engineering **3**(4) 1-36.

Kadlec, R.H. and R.L. Knight, 1996. Treatment Wetlands. Lewis Publishers, Boca Raton, FL, 893 pp.

King, D. and D.B. Nedwell, 1987. The adaptation of nitrate-reducing bacterial communities in esturaine sediments in response to overlying nitrate load. Microbiology Ecology **45** 15-20.

Knight, R.L., 1997. Wildlife habitat and public use benefits of treatment wetlands. Water Science and Technology **35**(5) 35-43.

Knight, R.L., R.A. Clarke, Jr. and R.K. Bastian, 2001. Surface flow (SF) treatment wetlands as habitat for wildlife and humans. Water Science and Technology **44**(11-12) 27-37.

Knowlton, M.F., C. Cuvellier and J.R. Jones, 2002. Initial performance of a high capacity surface-flow treatment wetland. Wetlands **22**(3) 522-527.

Kovacic, D.A., M.B. David, L.E. Gentry, K.M. Starks, and R.A. Cooke, 2000. Effectiveness of constructed wetlands in reducing nitrogen and phosphorus export from agricultural tile drainage. Journal of Environmental Quality **29** 1262-1274.

May, E., J.E. Butler, M.G. Ford, R. Ashworth, J.B. Williams and M.M.M. Bahgat, 1990. Chemical and microbiological processes in gravel-bed hydroponic (GBH) systems for sewage treatment, In: Constructed Wetlands in Water Pollution Control, edited by P. F. Cooper and B. C. Findlater, Pergamon Press, Oxford, UK, pp. 33-40.

Melzer, A. and D. Exler, 1982. Nitrate and nitrite reductase activities in aquatic macrophytes. In: Studies on Aquatic Vascular Plants, edited by J.J. Symoens, S.S. Hooper and P. Compère, Royal Botanical Society of Belgium, Brussels, pp. 128-135.

Metcalf & Eddy, Inc., 1991. Wastewater Engineering, Treatment, Disposal, and Reuse. Third Edition, revised by G. Tchobanoglous and F. L. Burton. McGraw-Hill, New York, NY.

Phipps, R.G. and W.G. Crumpton, 1994. Factors affecting nitrogen loss in experimental wetlands with different hydrologic loads. Ecological Engineering **3**(4) 399-408.

Spieles, D.J. and W.J. Mitsch, 2000. The effects of season and hydrologic and chemical loading on nitrate retention in constructed wetlands: a comparison of low- and high-nutrient riverine systems. Ecological Engineering **14**(1-2) 77-91.

State of Illinois, 2001. Statistical Loading Report, Permit Compliance System, Annual Point Source Loads for Ammonia (Illinois River).

Tanner, C.C., M.L. Nguyen, and J.P.S. Sukias, 2005. Nutrient removal by a constructed wetland treating subsurface drainage from grazed dairy pasture. Agriculture, Ecosystems and Environment **105** (1-2) 145-162.

Trepel, M., and L. Palmeri, 2002. Quantifying nitrogen retention in surface flow wetlands for environmental planning at the landscape scale. Ecological Engineering **19**(2) 127-140.

USEPA, 1999. Free Water Surface Wetlands for Wastewater Treatment: A Technology Assessment. USEPA 832-S-99-002, Dated June 1999, published November 2000, 165 pp.

Weisner, S.E.B., P.G. Eriksson, N.P., Graneli, and W.L. Leonardson, 1994. Influence of macrophytes on nitrate removal in wetlands. Ambio **23**(6)363-366.

Weller, M.W., 1978. Management of freshwater marshes for wildlife. In: Freshwater Wetlands: Ecological Processes and Management Potential, edited by R.E. Good, D.F. Whigham and R.L. Simpson, Academic Press, New York. pp. 267-284.

Westermann, P. and B.K. Ahring, 1987. Dynamics of methane production, sulfate reduction, and denitrification in a permanently waterlogged alder swamp. Applied and Environmental Microbiology **53**(10) 2554-2559.

Worrall, P., K.J. Peberdy and M.C. Millett, 1997. Constructed wetlands and nature conservation. Water Science and Technology **35**(5) 205-213

Constructed wetlands to retain contaminants and nutrients, specifically phosphorus from farmyard dirty water in Southeast Ireland

E.J. Dunne[1,2,3], N. Culleton[2], G. O'Donovan[3] and R. Harrington[4]
[1]*Wetland Biogeochemistry Laboratory, Soil and Water Science Department, University of Florida/IFAS, 106 Newell Hall, PO Box 110510, Gainesville, FL 32611, USA*
[2]*Teagasc Research Centre, Johnstown Castle, Wexford, Ireland*
[3]*Department of Environmental Resource Management, Faculty of Agriculture, University College Dublin, Belfield, Dublin 4, Ireland*
[4]*National Parks and Wildlife, Department of Environment, Heritage and Local Government, The Quay, Waterford, Ireland*

Abstract

Constructed wetlands, which are ecologically engineered systems, may be an appropriate method to help control pollution from agriculture. We investigated the ability of a farm-scale integrated constructed wetland (ICW) to treat incoming farmyard dirty water. We also determined the ability of two constructed wetland soils in Southeast Ireland to sorb and retain phosphorus (P) using small-scale studies (P adsorption isotherms and soil/water column studies). At the farm-scale, rainfall contributed most of the hydraulic inputs to the wetland, whereas most of the mass load of contaminants and nutrients were associated with farmyard dirty water. There was a significant relationship between farmyard dirty water generated and rainfall. There was no seasonal variability in the quality and/or quantity of farmyard dirty water generated at this particular site. Greater than 60% of the incoming mass load P (soluble reactive P) to the wetland was retained during the monitoring period. In the P adsorption isotherms, Dunhill constructed wetland soils had higher P sorption maximums (S_{max}) than Johnstown Castle site soils, which reflect differences in soil characteristics such as aluminium and iron content. Both soils had low equilibrium P concentrations (EPC_0) suggesting that these soils had a good capacity to sorb P. During the first two days of the soil/water column study, soils that were overlain by distilled water released P. After day two, soils typically retained P. Soils that had higher P concentrations in overlying waters (5 and 15 mg SRP l^{-1}) retained P at highest rates. This information is useful as it provides quantitative information at the operational and experimental wetland-scale that may be useful for future wetland design in Ireland.

Keywords: constructed wetlands, agriculture, contaminants, nutrients, phosphorus, farmyard dirty water.

Introduction

Eutrophication is a major threat to water quality in Ireland (Clenaghan *et al.*, 2001). Point source pollution from agriculture can include leakages from farmyard areas or the inappropriate management of farmyard dirty water (Healy *et al.*, 2003). In Ireland, dairy farmyard dirty water is commonly composed of farmyard runoff, parlour washings, silage and farmyard manure effluents along with general farmyard washings and in some cases yard roof waters.

Wetlands, both natural and constructed, are used to improve water quality within agricultural areas by typically intercepting and retaining incoming contaminants and nutrients through a series of vegetated ponds before waters leave the wetland system (Knight *et al.*, 2000). Constructed wetlands are ecologically engineered systems that are founded on basic ecological principles (Mitsch and Jørgensen, 1989). They are often used as alternatives to, or components of, conventional nutrient management practices to reduce or eliminate contaminant and nutrient loads in agricultural wastewaters around the world (Cronk, 1996; Peterson, 1998; Geary and Moore, 1999; Knight *et al.*, 2000; Borin *et al.*, 2001; Hunt and Poach, 2001).

In Ireland, the use of constructed wetlands to manage agricultural waters such as farmyard dirty water has been primarily based on a site-specific integrated approach. Integrated constructed wetlands (ICWs), which are a design-specific approach of conventional surface-flow constructed wetlands, were first used in the 1990s in a small agricultural watershed (Harrington and Ryder, 2002). An ICW is an integrated system, as it provides for water quality improvement and the presence of such a system within an agricultural landscape, also provides for additional ecological habitats.

The main objectives of this research were to: (1) determine the effectiveness of an operational farm-scale ICW to treat incoming farmyard dirty water; (2) physicohemically characterise two constructed wetland soils in the Southeast of Ireland; and (3) quantify phosphorus (P) sorption and retention capacities of these soils using small-scale experiments (P adsorption isotherms and soil/water column studies)

Material and methods

Site description

The farm-scale ICW was situated at the Teagasc Research Centre, Johnstown Castle, Wexford located in the Southeast of Ireland. Ireland in general has a cool temperate west maritime climate where annual rainfall is about 1,000 mm and mean temperature is 10°C (Gardiner and Radford, 1980). The wetland system was constructed in 2000 using in situ soils (complex of imperfectly drained gleys and well to moderate draining brown earths).

For this study farmyard dirty water, which was comprised of rainfall on open farmyard areas (2,031 m^2), farmyard manure and silage effluents, along with dairy and yard washings from a 42-cow organic dairy unit. It was collected in a central storage facility (three chambered concrete tank) prior discharge to the wetland.

The integrated constructed wetland was comprised of three surface flow treatment wetland cells with a total area of 4,265 m^2 and one final monitoring pond (490 m^2) (Figures 1 and 2). Up-gradient and surrounding the wetland system was unfertilised organic grassland. There was a 30 cm deep surface drain around the up gradient side of the wetland site. Wetland-

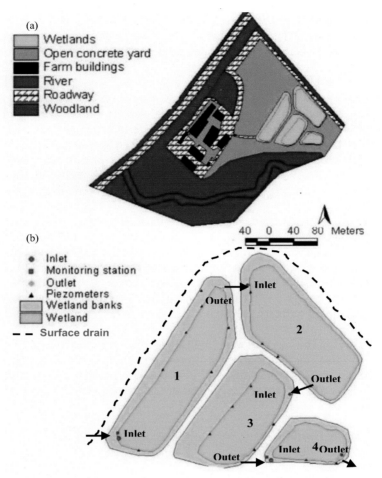

Figure 1. Sketch of (a) farmyard and (b) integrated constructed wetland layout with installed monitoring stations (Source: Dunne et al., *2005).*

treated waters were point discharged (whenever there was outflow) to adjacent riparian woodland which drained to a nearby stream.

Emergent wetland plant species plante covered about 80% of wetland surface area and included *Carex riparia* Curtis., *Typha latifolia* L., *Phragmites australis* (Cav.) Trin. ex Steudel, *Sparganium erectum* L., *Glyceria fluitans* (L.) R. Br., *Iris pseudacorus* L., *Phalaris arundinacea* L. and *Alisma plantago-aquatica* L.

Pre and during wetland monitoring, water quality monitoring stations were installed at the wetland inlet (inlet one) and wetland outlet (outlet of treatment wetland cell three) pipes in April 2001 and December 2001, respectively. Stations were equipped with portable water

Figure 2. Wetland cells one (1), two (2), three (3) and final monitoring pond (4) of the integrated constructed wetland at the Teagasc Research Centre, Wexford.

samplers. Pipe flows were measured on a continuous basis and during events a flow proportional sample was taken.

During wetland operation, water grab samples were taken fortnightly from wetland inlets two (I2), three (I3), and the outlet of the final monitoring pond (O4). Also, piezometer water samples were taken about every two months.

Daily rainfall and evaporation data were recorded at a local weather station (within 750 m of site). Evapotranspiration was estimated at 80% of Class A pan data (Kadlec and Knight, 1996).

Water quality analyses

Collected water samples from wetland inlets and outlets were stored at 4°C and analysed for five day biological demand (BOD_5) total suspended solids (TSS), and soluble reactive phosphorus (SRP) using standard water quality methods (APHA, 1992). Ammonium was measured colormetrically following Berthelot reduction method (Houba *et al.*, 1987).

Small-scale experiments

Soils were collected from the constructed wetlands at the Teagasc Research Centre, Johnstown Castle, Wexford and a private farm at Dunhill, Waterford (Figure 3) for two purposes. These were to: (1) physicochemically characterise site soils; (2) determine P sorption capacities of soils using P adsorption isotherms; and (3) determine P retention capacity of soils using soil/water column studies. At the Teagasc Research Centre site soils were classified as a complex of imperfectly drained gleys and well to moderate draining brown earths, whereas at the Dunhill site, soils were poorly drained alluvial soils.

Bulk- and depth-specific soil samples were taken at random from each site for soil characterisation. Soils were characterised for pH, particle size distribution, total nitrogen, total P, ammonium oxalate extractable P, iron (Fe) and aluminium (Al) using standard soil methods.

For P adsorption isotherms, one gram of dry, sieved and ground soil was incubated with 20 ml of 0.01 M KCl containing 0, 0.1, 0.5, 1, 5, 10, 50 and 100 mg P l^{-1}, as KH_2PO_4 in centrifuge tubes. Tubes were shaken, centrifuged and filtered to 0.45 µm. Filtrates were analysed for SRP. The sum of native P and P sorbed by soil represents the total amount of P adsorbed by the soil (Fresse *et al.*, 1992). The total amount of P adsorbed was then fitted to the Langmuir model (Rhue and Harris, 1999), which helps provide theoretical parameters such as maximum P sorption capacity of a soil (S_{max}) and equilibrium P concentration (EPC_0).

For the soil/water column study, 18 intact soil columns were randomly taken from each site. Distilled water was spiked at 0, 1, 5, and 15 mg P l^{-1} as KH_2PO_4. Spiked waters were then placed into columns that contained soils to a water depth of 15 cm. Soil/water columns were then incubated for 30 days. Each treatment was replicated three times. Overlying floodwater samples were taken regularly and analysed for SRP.

(1)   (2)

Figure 3. Integrated constructed wetland at (1) Teagasc Research Centre, Wexford and (2) private farm at Dunhill, Waterford.

Parameter determinations

A site-specific water balance equation was used for the farm-scale wetland at the Teagasc Research Centre (Equation 1).

$$\frac{dv}{dt} = Q_{in} + P_w + P_b - ET_w - Q_{out} \tag{1}$$

where Q_{in} is farmyard dirty water inflow, P_w is the precipitation on the wetland surface area (wetland cells one to three), P_b is inflow from precipitation on the surrounding wetland bank area, ET is the evapotranspiration from the wetland surface area (wetland cells one to three), and Q_{out} is the wetland surface outflow. All units are in cubic meters.

The wetland banks were grass covered, steep sloped and were isolated from surrounding surface hydrologic inflows. Soluble reactive P concentrations were not measured in surrounding wetland bank inflows to the wetland so literature values were used to allocate loads to contributing waters. Tunney *et al.* (1997) found that in a nearby site, maximum SRP concentrations in surface runoff from a small, low intensity grassland catchment was 0.054 mg l^{-1}.

For the soil water column study, total mass P load in the overlying water column was determined for each sampling period as outlined by Dunne (2004) (Equation 2).

$$(C_c * V_c) + (C_r * V_r) = C_t * V_t \tag{2}$$

where C_c is SRP concentration in water column, (mg l^{-1}); V_c is volume of overlying water in column, (l), which is the initial core volume (1.2 l) minus the overlying floodwater sample volumes taken at each sampling period. This mass did not take rainfall mass P inputs into account; C_r is SRP concentration of rainfall, (mg l^{-1}), which was a sampling period average; V_r is volume of rainfall onto core, (l); C_t is SRP concentration actually measured in overlying water column, (mg l^{-1}); and V_t is total volume (water in column, rainfall and floodwater sampling) of overlying water in column, (l). Phosphorus retained or released (mg m^{-2} d^{-1}) by soil was determined as the slope between cumulative P in overlying water column (mg m^{-2}) and days of incubation.

Results and discussion

Farm-scale integrated constructed wetland

Water balance
Using a site-specific water balance equation, all hydraulic inflows and outflows to and from the wetland were generally accounted for. This can be assumed as there was nearly a one to one relationship between the two parameters (Figure 4). Rainfall on wetland surface areas and surrounding wetland banks contributed about 73% of the hydraulic inputs to the wetland, while farm yard dirty water accounted for the remaining 27% during the study period.

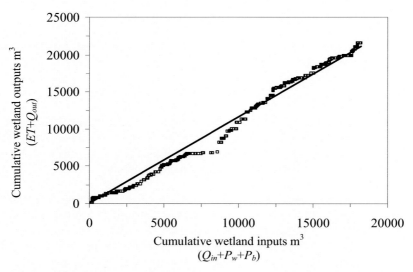

Figure 4. Relationship between cumulative hydraulic wetland inputs and outputs during the monitoring of the integrated constructed (Source: Dunne et al., 2005).

Hydraulic flows
Highest monthly rainfall was recorded in July 2001 (5.1 ± 2.9 mm d^{-1}) (mean ± standard error); October 2002 (8.8 ± 2.3 mm d^{-1}); and November 2002 (7.9 ± 1.4 mm d^{-1}). However, there was no significant seasonal variation in daily rainfall. There was a significant relationship between farmyard dirty water generated and rainfall during the monitoring period (Figure 5), which indicates that rainfall on impervious surfaces such as open yard areas may be important in generating dirty water. However, it was not a simple cause and effect relationship at this site with other site-specific factors needing consideration. These may include climatic factors and farm management practices (separation and storage of yard waters and effluents, and volumes of dairy wash water used). When there was inflow of farmyard dirty water to the farm-scale ICW, inflow rates varied between 4 and 19 m^3 d^{-1} (Figure 5). In general, there was little variation in volumes of farm yard dirty water generated on a seasonal basis at this particular site.

Water quality
There was no seasonal variability in the quality of farmyard dirty water that was discharged to the wetland. Average concentrations were 2806 ± 120 mg BOD$_5$ l^{-1}, 905 ± 43 mg TSS l^{-1}, 44.98 ± 2.29 mg NH$_4{}^+$-N l^{-1}, and 18.86 ± 0.87 mg SRP l^{-1}. These findings are contrary to other studies that investigated the characteristics of wastewaters generated in dairy farms in the U.K. (see Brewer *et al.*, 1999). In general, there were significant concentration reductions in water quality parameters between wetland inlets two, three and the outlet to the final monitoring pond. Concentrations of water quality parameters that were finally discharged to the final monitoring pond were relatively stable. Concentrations were 20 ± 3 mg BOD$_5$ l^{-1}, 11 ± 1 mg TSS l^{-1}, 2.0 ± 0.4 mg NH$_4{}^+$-N l^{-1}, and 1.7 ± 0.1 mg SRP l^{-1} (*n* = 140). During the monitoring period there were significant correlations between flow-weighted mean monthly

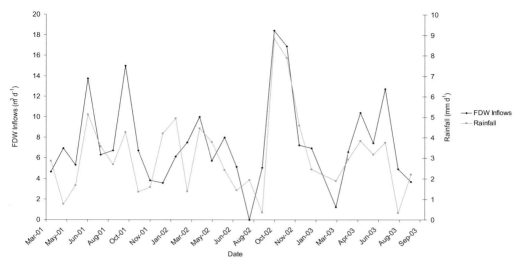

Figure 5. Farmyard dirty water wetland inflows and rainfall per month during the study (Source: Dunne et al., 2005).

farmyard dirty water parameter concentrations discharged to the wetland and outflows to the final monitoring pond (Table 1). Such relationships can help to generate empirical relationships, which may aide future farm-scale wetland design.

Mass loads

Mass inputs rates of BOD_5, TSS, NH_4^+-N, and SRP were somewhat variable, but this was not significant with season. Typically, mass inputs of BOD_5 and TSS, NH_4^+, and SRP from farm yard dirty water to the constructed wetland were less than the loading rates reported by similar studies (see Reaves *et al.*, 1994; Skarda *et al.*, 1994; Cronk, 1996; Newman *et al.*, 2000). Mass loading rates of organic and suspended material in farmyard dirty water to the ICW were 5,484

Table 1. Correlation matrix of mean monthly flow-weighted farmyard dirty water and outflow to the final monitoring pond during April 2001 - September 2003 (n = 16).

	Water quality parameter concentrations													
	SRP_{in}		NH_4^+-N_{in}		BOD_{5in}		TSS_{in}		SRP_{out}		NH_4^+-N_{out}		BOD_{5out}	
$NH_4^+{}_{in}$	0.561	*												
BOD_{5in}	0.032	NS‡	0.113	NS										
TSS_{in}	0.156	NS	0.172	NS	0.908	***								
SRP_{out}	0.610	**	0.182	NS	-0.351	NS	-0.234	NS						
$NH_4^+{}_{out}$	0.139	NS	0.184	NS	-0.260	NS	-0.304	NS	0.371	NS				
BOD_{5out}	0.044	NS	0.102	NS	0.587	*	0.452	NS	-0.080	NS	0.327	NS		
TSS_{out}	-0.030	NS	-0.237	NS	0.714	**	0.656	*	-0.404	NS	-0.178	NS	0.494	*

*, **, *** Significant at the 0.05, 0.01 and 0.001 probability levels, respectively

† SRP, soluble reactive P; BOD_5, five-day biological oxygen demand; and TSS, total suspended solids

‡ Not significant.

± 1433 kg yr^{-1}, and 1570 ± 465 kg yr^{-1}, respectively. Nutrient loading rates were 47 ± 10 kg SRP yr^{-1} and 128 ± 35 kg NH$_4^+$-N yr^{-1}. In terms of mass nutrient retention, SRP retention by the wetland system was only estimated, as P is typically the most land limiting nutrient. Thus, if one is retaining P, other material such as organic, suspended and N containing material is also being retained within the system. Total mass SRP load (the sum of SRP load in farmyard dirty water, precipitation on wetland surface areas and wetland bank inflow) to the wetland between December 2001 and September 2003 was 44 ± 11 kg SRP yr^{-1}, while output rates from the wetland were 15 ± 8 kg SRP yr^{-1} suggesting that retention rates during the study period was about 62%. Retention rates were variable with season with least amounts of SRP retained during winter periods, probably a result of increased wetland hydrological inputs during these periods.

Small-scale experiments

Phosphorus adsorption isotherms
Soils of both sites were generally acidic with pH values less than 6.7. Johnstown Castle site soils had lower total nitrogen and total P (1100 ± 29 mg kg^{-1} and 450 ± 29 mg kg^{-1}) than Dunhill constructed wetland soils (3917 ± 183 mg kg^{-1} and 867 ± 33 mg kg^{-1}). Also, Dunhill had highest clay (33 ± 2%) and silt (55 ± 2%) fractions. The differences in physicochemical characteristics suggests that Dunhill soils may have a higher capacity to retain P as Al and Fe oxides (which are known to control P sorption in mineral soils) are typically associated with clay fractions (Freese *et al.*, 1992).

In the batch experiments (phosphorus sorption isotherms) native adsorbed P was greater in Dunhill soils (935 ± 153 mg kg^{-1}) than in Johnstown Castle soils (156 ± 29 mg kg^{-1}). This is not surprising as Dunhill soils had the highest clay, TP, ammonium oxalate extractable Al (Al$_{ox}$), and Fe (Fe$_{ox}$) (2079 ± 194 mg kg^{-1} and 8013 ± 341 mg kg^{-1}).

The higher P sorption maximum values (S$_{max}$) as determined by the Langmuir model for Dunhill soils (1464 ± 157 mg kg^{-1}) in comparison to Johnstown Castle soils (618 ± 55 mg kg^{-1}) probably reflect differences in site soil characteristics such as Al$_{ox}$ and Fe$_{ox}$ content. Similarly, equilibrium P concentrations (EPC$_0$) were higher in Dunhill soils (17 µg l^{-1}) in comparison to Johnstown Castle soils (0.007 µg SRP l^{-1}). Equilibrium P concentration is the concentration at which soils neither release nor retain P; also, soil typically sorbs P at solution concentrations above EPC$_0$ values (Khalid *et al.*, 1977; Bridgham *et al.*, 2001). As both soils had low EPC$_0$, our findings suggest that these soils have a good capacity to sorb P. For this study, P sorption parameters could be predicted by Al$_{ox}$ and Fe$_{ox}$ content. Maximum P sorption was best predicted by Al$_{ox}$ (R^2 = 0.90; *p* < 0.001; *n* = 18), whereas EPC$_0$ was best predicted by Fe$_{ox}$ content (R^2 = 0.55; *p* < 0.001; *n* = 18) of constructed wetland soils. It is well established in the literature that P sorption is typically controlled by amorphous forms of Al and Fe in mineral soils (Faulkner and Richardson, 1989; Johnston, 1991).

Soil/water column study
During the first two days of the column study, soils that were overlain by distilled water released P from soil to water (Figure 6). After day two, these soils typically retained P and columns, which were initially spiked at 1, 5 and 15 mg SRP l^{-1} retained P during the 30 days

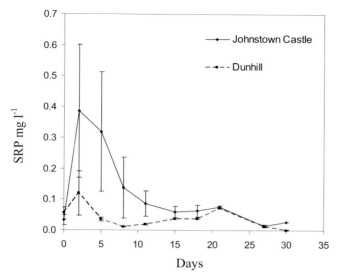

Figure 6. Changes in soluble reactive phosphorus (SRP) concentration in soils that were initially flooded with distilled water, during the 30 day study.

of the experiment (Figure 7). Numerous studies show that P release and retention from soils such as constructed wetland soils to overlying waters is dependent on the P concentration gradient between the two components (Reddy *et al.*, 1999; Pant and Reddy, 2003).

Phosphorus retention rates were similar between site soils with overlying floodwaters initially set at 0 mg and 1 mg SRP l^{-1} (1.0 and 1.5 mg m^{-2} d^{-1}). Soils that had higher P concentrations in overlying waters (initially spiked at 5 and 15 mg SRP l^{-1}) retained P at highest rates from overlying waters (17.4 ± 4.7 mg P m^{-2} d^{-1} and 55.9 ± 4.35 mg m^{-2} d^{-1}, respectively) and columns spiked at 15 mg SRP l^{-1} had highest retention rates. In addition, there were no site soil differences in retention rates at these higher concentrations. Several studies reported that as loading increases, the likelihood of retention also increases (Gale *et al.*, 1994; Reddy *et al.*, 1995).

Conclusions

Pollution from present agricultural practices can either directly or indirectly impact water quality, with eutrophication a major concern in Ireland. One management approach to help control pollution from agriculture could include the use of ecologically engineered systems such as constructed wetlands.

Results from our farm-scale monitoring study suggest that there was little seasonal variability in volumes and quality of farmyard dirty water produced. Farmyard dirty water discharged to an ICW, had reduced nutrient and contaminant concentrations, as well as reduced mass loads. The mass loads of SRP, NH_4^+-N, BOD_5, and TSS in farmyard dirty water prior wetland treatment, suggest that farm yard dirty water, if not managed appropriately, has a high polluting potential.

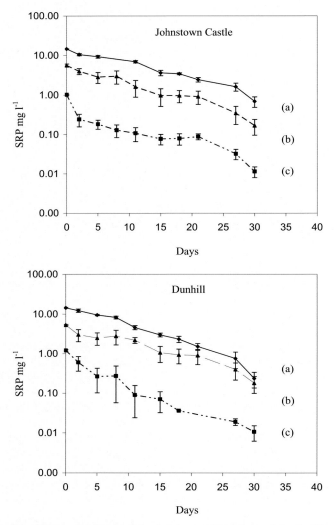

Figure 7. Changes in overlying water soluble reactive phosphorus (SRP) concentrations of soil/water columns of Johnstown Castle and Dunhill site soils. Overlying floodwaters of each column were initially spiked at (a) 15, (b) 5, and (c) 1 mg SRP l⁻¹. Error bars represent one standard error.

Precipitation had a significant effect on the amount of farmyard dirty water produced, but also on wetland performance to retain contaminants and nutrients, with the wetland typically retaining fewer nutrients, specifically P, during wet winter months.

Based on the results of our farm-scale study, future wetland design efforts in Ireland should consider extreme rainfall events so that appropriate wetland size and subsequently appropriate hydraulic retention times of wetland waters are achieved for successful water treatment.

From the small-scale experiments it was demonstrated that constructed wetland soils can retain P; however this is dependent on the concentration of overlying wetland water. Thus, when P concentrations in overlying waters are higher than underlying soils there is a net movement of P from overlying water to underlying soil and the reverse is true when concentrations are greater in underlying soils.

Findings of the P adsorption isotherms indicated that for both site soils investigated, P sorption was governed by amorphous non-crystalline forms of Fe and Al. It is important to note that P sorbed to Fe is typically released under anaerobic conditions in constructed and/or natural wetland soils, whereas P sorbed by Al is not affected by changes in redox conditions. Several wetland studies have suggested that anaerobic wetland soils can sorb about 70% of the P sorbed under aerobic conditions (Gale *et al.*, 1994; Reddy *et al.*, 1998).

In conclusion, small-scale studies such as those undertaken are useful in providing baseline information which may be useful for future wetland design and decision making purposes. Thus, it is important to "scale up" to large-scale ecosystem studies such as the farm-scale operational wetland as we did, since ecosystem components such as soil, water, vegetation, and biological organisms and their interactions are not easily simulated in small-scale studies. In an integrated approach with experimentation at several scales, a better understanding of ecosystem dynamics may be gained, such that information based on sound science can be provided to help formulate management decisions.

Acknowledgements

This research was funded by the Irish Teagasc Walsh Fellowship scheme. The authors acknowledge the technical help of Teagasc Research Centre, Johnstown Castle staff. The authors acknowledge the help of Mr. Arne Olsen for help with the water balance equation.

References

APHA, 1992. Standard methods for the examination of water and wastewaters. 18th Edition. American Public Health Association. Washington, D.C.

Borin, M., G. Bonaiti, G. Santamaria, and L. Giardini, 2001. A constructed surface flow wetland for treating agricultural waste waters. Water Science and Technology **44** 523-530.

Cleneghan, C., 2003. Phosphorus regulations: national implementation report, 2003. EPA. Wexford, Ireland.

Cronk, J.K., 1996. Constructed wetlands to treat wastewater from dairy and swine operations: a review. Agriculture Ecosystems and Environment. **58** 97-114.

Dunne, E.J., 2004. Wetland systems to mitigate contaminant and nutrient loss from agriculture. Ph.D. Dissertation, University College Dublin, Ireland.

Dunne, E.J., N. Culleton, G. O'Donovan, R. Harrington, A.E. Olsen, 2005. An integrated constructed wetland to treat contaminants and nutrients from dairy farmyard dirty water. Ecological Engineering 24 221-234.

Freese, D., S.E.A.T.M. van der Zee, and W.H. van Riemsdijk, 1992. Comparison of different models for phosphate sorption as a function of the iron and aluminium oxides of soils. Journal of Soil Science **43** 729-738.

Gale, P.M., K.R. Reddy and D.A. Graetz, 1994. Wetlands and aquatic processes: phosphorus retention by wetland soils used for treated wastewater disposal. Journal of Environmental Quality **23** 370-377.

Gardiner, M.J. and T. Radford, 1980. Soil associations of Ireland and their land use potential: explanatory bulletin to soil map of Ireland 1980. An Foras Talúntais, Dublin.

Geary, P.M., and J.A. Moore, 1999. Suitability of a treatment wetland for dairy wastewaters. Water Science and Technology **40** 179-185.

Harrington, R., and C. Ryder, 2002. The use of integrated constructed wetlands in the management of farmyard runoff and waste water. In: Proceedings of the National Hydrology Seminar on Water Resource Management: Sustainable Supply and Demand. Tullamore, Offaly. 19th Nov. 2002. The Irish National Committees of the IHP and ICID, Ireland.

Healy, M.G., M. Rodgers, and J. Mulqueen, 2004. Recirculating sand filters for treatment of synthetic dairy parlor washings. Journal of Environmental Quality **33** 713-718.

Houba, V.J.G., I. Novozamsky, J. Uittenbogaard and J.J., van der Lee, 1987. Automatic determination of total soluble nitrogen in soil extracts. Landwirtschaftliche Forschung. **40** 295-302.

Hunt, P.G. and M.E. Poach, 2001. State of the art for animal wastewater treatment in constructed wetlands. Water Science and Technology **44** 19-25.

Kadlec, R.H. and R.L. Knight, 1996. Treatment wetlands. Lewis Publishers, Boca Raton. 893 pp.

Knight, R.L., V.W.E. Payne Jr., R.E. Borer, R.A. Clarke Jr., and J.H. Pries. 2000. Constructed wetlands for livestock wastewater management. Ecological Engineering **15** 41-55.

Mitsch, W.J. and J.G. Gosselink, 1993. Wetlands. 2nd Edition. Jon Wiley and Sons, Inc., New York. 722 pp.

Mitsch, W.J., and S.E. Jørgensen, 1989. Classification and examples of ecological engineering. pp. 12-19. In: Ecological engineering: an introduction ecotechnology, edited by Mitsch, W.J., and S.E. Jørgensen, Jon Wiley and Sons, New York. 411 pp.

Newman, J.M., J.C. Clausen, and J.A. Neafsey, 2000. Seasonal performance of a wetland constructed to process dairy milkhouse wastewater in Connecticut. Ecological Engineering **14** 181-198.

Peterson, H.G. 1998. Use of constructed wetlands to process agricultural wastewater. Canadian Journal of Plant Science **78** 199-210.

Reaves, R.P. P.J. DuBowy, and B.K. Miller, 1994. Performance of a constructed wetland for dairy waste treatment in LaGrange County, Indiana. In: Proceedings of a Workshop on Constructed Wetlands for Animal Waste Management, edited by P.J. DuBowy and R.P. Reaves, 4-6 April, 1994, Lafayette, IN. pp. 43-52.

Reddy, K.R., G.A. O Connor, and P.M. Gale, 1998. Phosphorus sorption capacities of wetland soils and stream sediments impacted by dairy effluent. Journal of Environmental Quality **27** 438-447.

Rhue, R.D. and R.G. Harris, 1999. Phosphorus sorption/desorption reactions in soils and sediments. pp.187-206. In: Phosphorus biogeochemistry in subtropical ecosystems, edited by Reddy, K.R., G.A. O'Connor, and C.L. Schleske, Lewis Publishers, Boca Raton, 707 pp.

Skarda, S.A., J.A. Moore, S.F. Niswander, and M.J. Gamroth, 1994. Preliminary results of wetland for treatment of dairy farm wastewater. In: Proceedings of a Workshop on Constructed Wetlands for Animal Waste Management, edited by P.J. DuBowy and R.P. Reaves, 4-6 April, 1994, Lafayette, IN., pp. 34-42.

Tunney, H., T. O'Donnell, and A. Fanning, 1997. Phosphorus loss to water from a small low intensity grassland catchment. pp. 358-361. In: Phosphorus loss from soil to water, edited by Tunney, H., Carton, O.T., Brookes, P.C., and Johnston, A.E. CAB (Centre of agriculture and biosciences) International. Oxon. 352 pp.

Werner, T.M., and R.H. Kadlec, 2000. Wetland residence time distribution modeling. Ecological Engineering **15** 77-90.

The use of constructed wetlands with horizontal sub-surface flow for treatment of agricultural drainage waters together with sewage

J. Vymazal
ENKI o.p.s., Dukelská 145, 379 01 Trebon, Czech Republic

Abstract

Constructed wetlands with horizontal sub-surface flow (HF CWs) provide suitable conditions for removal of nitrate via denitrification due to anoxic or anaerobic conditions in wetland bed substrates. Agricultural field drainage usually contains high concentrations of nitrate, but these waters are very often low in dissolved organics necessary for denitrification. The combination of municipal sewage high in organics and agriculture drainage waters is a possible solution for treatment of both types of wastewater in one constructed wetland. However, this approach has been seldom used. In the Czech Republic, there are two HF CWs that have been designed to treat such a mixture. The results from both constructed wetlands indicate high removal of BOD_5 (with outflow concentrations less than 5 mg l^{-1}) and good removal of nitrate (outflow concentrations less than 5 mg l^{-1}; removal effect greater than 70%). On the other hand, removal of ammonia is very low (< 30%) and therefore, some further oxidation steps are required to increase effectiveness of the systems.

Keywords: agriculture drainage, constructed wetland, nitrate, sewage.

Introduction

Constructed wetlands with horizontal sub-surface flow provide suitable conditions for removal of nitrate via denitrification, as the filtration bed is usually anoxic or anaerobic (Kadlec and Knight, 1996; Vymazal, 1999, 2001a). At present, the majority of HSF CWs used to treat domestic or municipal sewage, which typically contain low concentrations of nitrate, while organic-N and ammonia-N are the predominant forms of nitrogen (Kadlec and Knight, 1996, Vymazal *et al.*, 1998). Agricultural field drainage usually contains high concentrations of nitrate, but low concentrations of dissolved organics necessary for denitrification. The combination of municipal sewage high in organics and agriculture drainage waters high in nitrate is a possible solution for treatment of both types of wastewater in one type of constructed wetland. However, this approach has been used very seldom so far.

Study sites

In the Czech Republic, there are three constructed wetlands where agricultural drainage is intentionally brought to the village sewer system. The systems at Chmelná and Onsov have been in operation for more than 10 years, the system at Brehov is quite new and there are no sufficient data available at present. The basic design parameters of constructed wetlands at Chmelná and Onsov are given in Table 1. At Chmelná, pre-treatment consists of a horizontal

Table 1. Design parameters of constructed wetlands at Chmelná and Onsov.

	Start of operation	PE	Size (m²)	Substrate	Fraction (mm)	Flow (m³ d⁻¹)	HLR (cm d⁻¹)	Plants
Chmelná	1992	150	706	crushed rock	5 - 10	90.0	12.7	*Phalaris sp.*
Onsov	1993	420	2100	gravel	3 - 6	103	4.9	*Phragmites sp.*

grit chamber and small Imhoff tank. Both systems are equipped with a stormwater overflow. Systems were monitored on a monthly basis during the period of 1994-1995 and bimonthly in 1996.

Results and discussion

The combination of sewage, agricultural drainage water and stormwater runoff had a very low concentration of organics in inflows (Figure 1). The mean concentrations for a three year period were only 19.7 mg l^{-1} and 27.4 mg l^{-1} at Chmelná and Onsov, respectively. Also, the respective inflow loads of organics (BOD_5) to vegetated beds were only 14.1 kg ha^{-1} d^{-1} and 5.4 kg ha^{-1} d^{-1}, which are well below the average value of 31.8 kg ha^{-1} d^{-1} for the Czech constructed wetlands (Vymazal, 2001b). The outflow BOD_5 concentrations were as low as 1.8 mg l^{-1} at Chmelná and 4.8 mg l^{-1} at Onsov, resulting in respective removal efficiencies of 91% and 78%.

Due to the input of drainage waters, inflow nitrate-N concentrations were quite high about 18.5 mg l^{-1} at Chmelná and 9.6 mg l^{-1} at Onsov (Figure 1) and the removal of nitrate-N amounted to 74% and 78% at Chmelná and Onsov, respectively. Despite theoretically suitable conditions for nitrate removal in HF CWs, high removal is not always achieved, probably because of partial oxygenation of the filtration bed (Vymazal 2001a). It is possible that the complete denitrification was limited by lack of organic material, especially at Chmelná, where outflow BOD_5 concentration was less than 2 mg l^{-1}. Kadlec and Knight (1996) reported that about 0.93 mg of organic carbon is necessary for denitrification of 1 mg of nitrate nitrogen.

The removal of ammonia was low and reached only 22% at Chmelná and 30% at Onsov. This was expected, but the removal was somewhat lower, as compared to an average ammonia removal of 39% for 151 HF CWs in Europe and North America and mean ammonia removal of 45% for 32 Czech HF CWs reported by Vymazal (2001a). But mean inflow ammonia concentrations at Chmelná and Onsov were quite low (Figure 1) and it is well known that removal efficiency increases with increasing inflow concentration. The data indicate that the filtration bed is predominantly anoxic/anaerobic, but aerobic nitrification does proceed to some extent in the wetland. The inflow total nitrogen (TN) concentrations were 33.6 mg l^{-1} and 20.7 mg l^{-1} at Chmelná and Onsov, respectively and removal efficiency amounted to 55% at Chmelná and 41% at Onsov, which is comparable to treatment effects reported elsewhere (Vymazal, 2001a). Despite low inflow concentrations, the average loading of the wetland at Chmelná was very high (1455 g N m² yr⁻¹) due to high hydraulic loading rate (12.6 cm d⁻¹), which is about twice higher than rates usually applied for sewage treatment in HF CWs (Vymazal, 2001a). The high flow was a consequence of a substantial contribution of

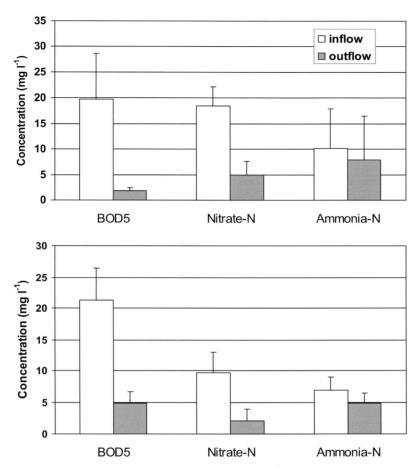

Figure 1. Removal of BOD$_5$, nitrate and ammonia in constructed wetlands at Chmelná (top) and Onsov (bottom). Results are means for the period 1994-1996.

agricultural drainage waters. Loading at Onsov was lower (316 g N m^2 yr^{-1}) due to much lower contribution of agricultural waters.

Due to high degree of dilution of sewage with waters of low ammonia concentrations (stormwater runoff and agricultural drainage), the outflow concentrations of ammonia were quite low (Figure 1). In addition there was low removal of ammonia in the system. There are no discharge limits for ammonia in treatment systems less than 2000 population equivalents (PE) in the Czech Republic and therefore, ammonia is usually not a primary target. It seems that single HF CWs are not able to reduce both nitrate and ammonia and for higher ammonia, additional treatment would be required. Constructed wetlands such as a vertical flow bed may offer an additional approach.

Conclusions

Treatment of agricultural drainage together with sewage and stormwater runoff in horizontal sub-surface flow is effective for removal of organics and nitrate nitrogen. However, removal of ammonia nitrogen is very low and to remove more ammonia from incoming wastewaters an additional oxidation step would be required.

Acknowledgements

The study was supported by grant MSM 000020001 "Solar Energetics of Natural and Technological Systems" from the Ministry of Education and Youth of the Czech Republic and by grant No. 206/02/1036 "Processes Determining Mass Balance in Overloaded Wetlands" from the Grant Agency of the Czech Republic.

References

Kadlec, R.H. and R.L. Knight, 1996. Treatment Wetlands. CRC/Lewis Publishers, Boca Raton, FL, USA, 893 pp.

Vymazal, J., 1999. Nitrogen removal in constructed wetlands with horizontal sub-surface flow - can we determine the key process? In :Nutrient Cycling and Retention in Natural and Constructed Wetlands, edited by J. Vymazal, Backhuys Publishers, Leiden, The Netherlands, pp. 1-17.

Vymazal, J., 2001a. Types of constructed wetlands for wastewater treatment: their potential for nutrient removal. In: Transformations on Nutrients in Natural and Constructed Wetlands, edited by J. Vymazal, Backhuys Publishers, Leiden, The Netherlands, pp. 1-93.

Vymazal, J., 2001b. Removal of organics in Czech constructed wetlands with horizontal sub-surface flow. In: Transformations on Nutrients in Natural and Constructed Wetlands, edited by J. Vymazal, Backhuys Publishers, Leiden, The Netherlands, pp. 305-327.

Vymazal, J., H. Brix, P.F. Cooper, M. Green and R. Haberl, (editors), 1998. Constructed wetlands for wastewater treatment in Europe. Backhuys Publishers, Leiden, The Netherlands, 348 pp.

Wetlands for treatment of nutrient laden runoff from agricultural lands: models and stormwater treatment area 1W, as a case study for protection of the Everglades ecosystem, USA

J.R. White and M.A. Belmont
Wetland Biogeochemistry Institute, Louisiana State University, Baton Rouge, LA 70803, USA

Introduction

Both natural and constructed wetlands are utilised around the world to intercept, collect and treat wastewater and stormwater runoff from the uplands prior discharge to adjacent water bodies (e.g. rivers, lakes, and estuaries) (Mitsch, 1994; Kadlec and Knight, 1996). We propose that constructed wetland placement within an agricultural watershed to treat agricultural runoff can follow two general models. Model I involves the construction of small treatment systems on the site of each individual farm/agricultural operations (Figure 1). This model requires the individual landowner to set aside some land area to build the wetland, as well as providing any infrastructure (canals, and swales) to direct the runoff to the wetland. This model is best suited for situations when agricultural operations directly flank the aquatic resource that is being considered for protection, (e.g. an adjacent stream, which runs throughout the landscape or lake surrounded by agriculture). The advantage of this model is that it allows some flexibility in wetland design and placement (Belmont *et al.*, 2004), but it is difficult to manage on a watershed-scale with many individuals responsible for site management, operation and considerable labour involved with potential compliance checks.

Model II involves the construction of a single wetland at the bottom of a watershed, which would capture all the surface water runoff that moves through the watershed (Figure 1). Therefore, a community effort would be potentially required to set aside or purchase an appropriate parcel of land, which would be topographically advantageous to surface water movement/drainage within the watershed. Canal construction may be required to direct surface water towards the constructed wetland. This model is best suited when the aquatic resource of concern is located at the lower part of the watershed. The advantages of this model is that the wetland is under one management scheme, requires one compliance location (outflow) and would generally have some support by the community as project development would likely involve several stakeholders that may include local and state regulatory agencies. The disadvantages include potentially converting a substantial land parcel out of agricultural production, perhaps requiring relocation of businesses, houses and rerouting of traffic thruways and some continuing costs in the way of community resources to manage/maintain the system. However, water quality regulations are becoming increasingly stringent; thereby necessitating alternative land uses to meet those criteria.

While these two models may be applicable in certain circumstances, many watersheds are likely best suited for a "mixed model" approach. For example, several landowners would come together

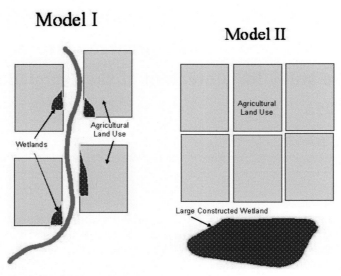

Figure 1. Models for constructed wetland placement within the watershed.

to manage surface runoff within their sub-basin of the watershed. Diversion of the surface water runoff could be implemented, while still maintaining a relatively small wetland footprint. The treated surface water could either be discharged into a canal, ditch or directly into the aquatic resource.

Diverting runoff from several farms into a small, collective constructed wetland helps reduce the total number of wetlands needing construction, management, and monitoring as compared to Model I. Again, the location of the individual agricultural operations to streams and rivers in the watershed will dictate the extent to which, larger and more collective systems can be maintained.

The Everglades Agricultural Area (EAA) is >200,000 ha of primarily agricultural land located in South-central Florida, USA just south of Lake Okeechobee and directly north of the Florida Everglades (Figure 2). Historically, runoff from agricultural fields was diverted into canals, which drain into the northern portion of the Everglades. This led to nutrient impaction, primarily by phosphorus (P), of this ecosystem (Reddy *et al.*, 1993). The increase in nutrient loading with time has led to a number of disturbances to ecological processes such as changes in native vegetation communities (Davis, 1991; Wu *et al.*, 1997) and nutrient cycling (DeBusk and Reddy, 1998; White and Reddy, 2000, 2003).

The state of Florida and the US Federal Government, acting as partners, have sought to establish large constructed wetlands at the edge of this extensive agricultural tract to intercept and treat agricultural runoff prior to discharge to the Everglades. Due to the size of the agricultural area and the proximity of the Everglades, it was decided that the placement of constructed wetlands would follow Model I. Agricultural land was purchased and converted to about 16,000 ha of wetlands or stormwater treatment areas (STAs) separated into six different systems. The STAs are required to function at a very low level of nutrient concentrations, allowing no more

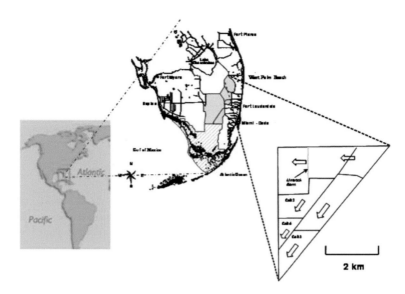

Figure 2. Location of Stormwater Treatment Area 1-West located north of the Everglades.

than 0.050 mg TP l⁻¹ to leave the STA, due to the extreme sensitivity of the receiving water body (Everglades ecosystem). The average inflow concentrations are about 0.2 mg TP l⁻¹.

Stormwater Treatment Area 1West is one of these large constructed wetlands and is 1,600 ha in size (Figure 3). Cell 5 of the STA-1W is a recent addition to the STA network at 800 ha, containing primarily submerged aquatic vegetation (SAV). The objective of our study was to sample surface water of this system to determine the nutrient removal capacity of this newly constructed STA.

Materials and methods

Water samples were collected, acidified and/or filtered as required and then placed on ice for transport to the laboratory (APHA, 1998). Analyses for P included soluble reactive P (SRP), total dissolved P (TDP) and total P (TP), which were determined colorimetrically (USEPA, 1993). Nitrogen species measured included ammonium, nitrate, nitrite, and total nitrogen (USEPA, 1993). Total suspended solids were determined by filtration using glass fiber filters (APHA, 1998). Dissolved oxygen (DO) was measured with a YSI meter (APHA, 1998).

Results and discussion

The results for water analyses were averaged for the influent (14 stations), east side of the limerock berm (5 stations), west side of the limerock berm (7 stations) and effluent (12 stations) (Table 1; Figure 4). A comparison of the average concentrations of the stations adjacent to both sides of the limerock berm was done to determine if the placement of the berm is benefiting P removal.

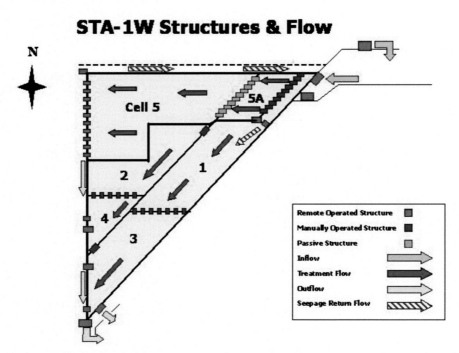

Figure 3. Schematic of Stormwater Treatment Area 1-W showing inflow, outflow, and water control structures (after Newman et al., *2003).*

During the study, mean SRP concentrations were reduced from 0.055 in the influent to 0.012 mg P l^{-1} in the effluent, TDP from 0.080 to 0.029 mg P l^{-1}, and TP from 0.120 to 0.040 mg P l^{-1} (Table 1). These changes correspond to concentration reductions of 78%, 62% and 67%, respectively. Therefore, Cell 5 of STA-1W is reducing P concentrations effectively. The P concentrations were not significantly different between the east and west side of the limerock berm, which suggests that the berm is not affecting P concentration in the water column. In general, STA effluent concentrations were below their design criteria of 0.050 mg TP l^{-1}.

Ammonium concentration averaged 0.295 mg N l^{-1} in the influent and 0.076 mg N l^{-1} in the effluent, while nitrate+nitrite average concentrations were 0.252 mg N l^{-1} in the influent and 0.089 mg N l^{-1} in the effluent (Table 1). Therefore, inorganic N removal was 74 and 64% for ammonium and nitrate+nitrite, respectively. The removal of total nitrogen in the wetland was low, with a total nitrogen (TN) concentration of 3.38 mg N l^{-1} in the influent reduced to 2.79 mg N l^{-1} in the effluent (17% reduction).

Total suspended solids were reduced in the wetland by 73%, from an average concentration of 8.91 to 2.66 mg l^{-1} at effluent monitoring stations (Table 1). Dissolved oxygen (DO) was increased by the SAV from an average concentration of 2.81 mg l^{-1} in the influent to 6.84 mg l^{-1} in the effluent. The vegetation coverage observed corresponded to 80% of *Hydrilla verticillata* and the balance of *Najas guadalupensis*, with traces of *Ceratophylum demersum*.

Table 1. *Summary of water analyses. All concentrations of soluble reactive phosphorus, (SRP); total dissolved phosphorus, (TDP); total phosphorus, (TP); ammonium, (NH_4^+); nitrate and nitrite, $(NO_3^-$ and $NO_2^-)$; total nitrogen, (TN); total suspended solids, (TSS); and dissolved oxygen, (DO) are reported in mg l^{-1}. Averages are for the influent, east side of the limerock berm, west side of the limerock berm and effluent of the STA system. The numbers in parentheses are one standard error.*

Parameter	Sampling Date	Influent	East side of limerock berm	West side of limerock berm	Effluent
SRP	08/21/03	0.082(0.010)	0.053(0.016)	0.047(0.009)	0.018(0.005)
	01/5/04	0.028(0.005)	0.018(0.005)	0.017(0.003)	0.006(0.002)
TDP	08/21/03	0.105(0.007)	0.077(0.018)	0.070(0.011)	0.034(0.006)
	01/5/04	0.055(0.006)	0.038(0.006)	0.033(0.003)	0.024(0.005)
TP	08/21/03	0.101(0.021)	0.102(0.037)	0.051(0.007)	0.034(0.006)
	01/5/04	0.138(0.007)	0.115(0.015)	0.097(0.022)	0.046(0.007)
NH_4^+	08/21/03	3.907(0.116)	3.346(0.077)	3.563(0.125)	3.208(0.084)
	01/5/04	2.751(0.235)	2.784(0.226)	2.253(0.101)	2.381(0.180)
NO_3^- / NO_2^-	08/21/03	0.483(0.071)	0.161(0.042)	0.298(0.061)	0.161(0.047)
	01/5/04	0.021(0.002)	0.051(0.010)	0.038(0.007)	0.017(0.003)
TN	08/21/03	3.907(0.116)	3.346(0.077)	3.563(0.125)	3.208(0.084)
	01/5/04	2.751(0.235)	2.784(0.226)	2.253(0.101)	2.381(0.180)
TSS	08/21/03	6.00(1.03)	11.60(5.18)	5.17(1.45)	1.00(0.43)
	01/5/04	11.82(4.92)	12.27(7.30)	6.00(2.74)	4.31(1.28)
DO	08/21/03	2.18(0.47)	4.97(2.44)	2.77(0.55)	6.49(1.26)
	01/5/04	3.43(0.66)	8.18(3.13)	5.63(1.29)	7.19(0.78)

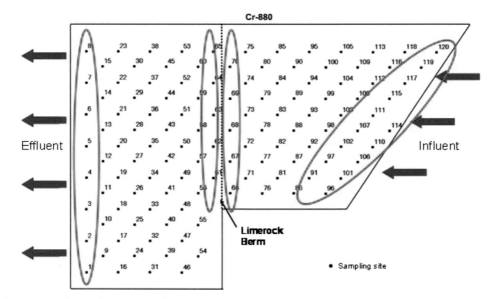

Figure 4. *Sampling stations in cell 5 of STA-1W. Water quality data were averaged within the circled stations.*

While this wetland is functioning well in terms of reduction of P, the wetland is currently in the start-up phase (within the first two years). With time, the treatment effectiveness may vary. However, continual adaptive management of the system that may include hydrology and vegetation management may help to sustain and perhaps improve the long term P removal capacity (White *et al.*, 2004).

Conclusions

Cell 5 of STA-1W is functioning well in reducing total P to 0.040 mg l^{-1}. Total suspended solids were reduced over 70%. Inorganic N was reduced by 69%, while TN concentrations were reduced by 17%. The DO levels in the surface water increased appreciably with a mean effluent concentration of greater than 6 mg O_2 l^{-1}. Currently, cell 5 of STA-1W is performing better than originally designed in removing P to help protect the Everglades ecosystem. Continuous monitoring and adaptive management will be required to improve the TP removal capacity down to 0.010 mg l^{-1} which is the background concentration of P within the Everglades national park.

References

APHA, AWWA, WPCF, 1998. Standard Methods for the Examination of Water and Wastewater. 20th edition, APHA, Washington DC, USA.

Belmont, M.A., E. Cantellano, S. Thompson, M. Williamson, A. Sánchez and C.D. Metcalfe, 2004. Treatment of domestic wastewater in a pilot-scale natural treatment system in Central Mexico. Ecological Engineering **23** 299-311.

Davis, S.M., 1991. Growth, decomposition and nutrient retention of *Cladium jamaicense*, Cranz. and *Typha domingensis* Pers. in the Florida Everglades. Aquatic Botany **40** 203-224.

DeBusk, W.F and K.R. Reddy, 1998. Turnover of detrital organic carbon in a nutrient-impacted Everglades marsh. Soil Science Society of America Journal **62** 1460-1468.

Kadlec, R.H. and R.L. Knight, 1996. Treatment Wetlands. CRC/Lewis Publishers, Boca Raton, FL. USA, 893 pp.

Mitsch, W.J. 1994. The non-point source pollution control function of natural and constructed wetlands. In: Global Wetlands: Old World and New, edited by Mitsch, W.J., Elsevier Science B.V. Amsterdam, pp. 351-361.

Newman, J.M., G. Goforth, M.J. Chimney, J. Jorge, T. Bechtel, G. Germain, M.K. Nungesser, D. Rumbold, J. Lopez, L. Fink, B. Gu, K. Cayse, R. Bearzotti, D. Campbell, C. Combs, K. Pietro, N. Iricanin, R. Meeker, G. West and N. Larson, 2003. Chapter 4: Stormwater Treatment Areas, SFWMD Everglades Consolidated Report. South Florida Water Management District, West Palm Beach, FL., pp. 4A-1 to 4C-76.

Reddy, K.R., R.D. Delaune, W.F. DeBusk and M.S. Koch, 1993. Long-term nutrient accumulation rates in the Everglades. Soil Science Society of America Journal **57** 1147-1155.

US Environmental Protection Agency, 1993. Methods for Chemical Analysis of Water and Wastes. EPA-600/4-79-020

White J.R. and K.R. Reddy, 2000. Influence of phosphorus loading on organic nitrogen mineralization of Everglades Soils. Soil Science Society of America Journal **64** 1525-1534.

White J.R. and K.R. Reddy, 2003. Nitrification and denitrification rates of Everglades wetland soils along a phosphorus-impacted gradient. Journal of Environmental Quality **32** 2436-2443.

White, J.R. K.R. Reddy and M.Z. Moustafa, 2004. Influence of hydrologic regime and vegetation on phosphorus retention in Everglades stormwater treatment area wetlands. Hydrological Processes **18** 343-355.

Wu, Y., F.S. Sklar, and K.R. Rutchey, 1997. Analysis and simulations of fragmentation patterns in the Everglades. Ecological Applications **7** 268-276.

Treatment wetlands for removing phosphorus from agricultural drainage waters

T.A. DeBusk, K.A. Grace and F.E. Dierberg
DB Environmental, Inc., 365 Gus Hipp Blvd., Rockledge, FL 32955, USA

Abstract

As a treatment technology, wetlands face several challenges in providing effective phosphorus (P) removal from agricultural drainage waters (ADW). Wetland area requirements for P removal are typically higher than other ADW constituents, such as nitrate-nitrogen and oxygen demanding substances. Moreover, P cycling within wetlands is complex, with exchanges between dissolved and particulate P forms, and labile and refractory P forms, occurring in the treatment wetland on a spatial and temporal basis. The gradual accumulation of P-enriched sediments with time can affect biogeochemical P removal pathways and limit long-term P removal effectiveness of treatment wetlands. Despite these challenges, wetlands are capable of reducing P in ADWs to extremely low levels, in the range of 15 - 20 μg l^{-1}. However, because such low outflow concentrations are only attained at low mass P loading rates (< 1 - 2 g P m^{-2} yr^{-1}), wetland area requirements per unit mass of P removal can be extremely high. Unit area requirements appear to decline under higher mass P loading conditions, but this is achieved at the expense of higher outflow P concentrations. Several techniques have been evaluated for improving wetland P removal effectiveness and sustainability, including routine vegetation harvest, removal of accumulated sediments, and chemical immobilization of P in sediments. Such practices have been shown to work in pilot-scale systems, but their technical and economic feasibility for full-scale use remains to be demonstrated.

Keywords: phosphorus cycling, sediment accumulation, phosphorus biogeochemistry.

Introduction

In the past two decades, constructed wetlands have become increasingly popular as a technology for removing nutrients from point and non-point source flows (Reddy and Smith, 1987; Kadlec and Knight, 1996). A suite of compounds, including oxygen demanding substances and nitrogen, are removed effectively and sustainably by treatment wetlands (Kadlec and Knight, 1996) due to the complete or partial conversion of these constituents to gaseous forms. Phosphorus (P) removal in wetlands has proven to be particularly challenging, because soils are the only long-term sink for this element (DeBusk and DeBusk, 2000; DeBusk and Dierberg, 1999). In order to improve long-term P removal performance by constructed wetlands, a number of design approaches and management practices have been investigated, with varying levels of success.

In this paper, we discuss P cycling processes in agricultural watersheds and treatment wetlands, and review findings of some previous studies on the use of constructed wetlands for treating agricultural drainage waters (ADWs). Finally, we discuss some of the design and management aspects of wetlands that may improve their performance and sustainability of P removal.

Phosphorus forms and wetland removal processes

Particulate phosphorus

Many agricultural runoff and wastewater streams contain high concentrations of suspended solids. Soil erosion is a prominent source of such particulate matter, with export of particulate P most common during periods of heavy rainfall and/or irrigation in agricultural areas. There are a number of Best Management Practices (BMPs) that can be implemented for minimising soil erosion, and hence, P losses from agricultural fields. Despite on-farm control efforts, some particulate P is inevitably present in ADWs, and this fraction often comprises the dominant P form.

Particles exported in drainage waters contain P of varying concentrations, with the P content related to the origin of the particles. Sharpley (1999) noted that manures for pigs, sheep and cattle range from 6700 - 17,600 mg P kg^{-1}, which is markedly higher than the P content of most non-impacted soils. Hence, particulate-laden runoff from animal husbandry operations can exhibit very high TP concentrations. Autochthonous particulate P generation can also occur within water conveyance structures in agricultural regions. For example, particulate detritus from floating macrophytes that proliferated in south Florida agricultural drainage canals contained an average of 4227 mg P kg^{-1} (Stuck, 1996)

The tendency for particulate matter to be mobilised from farm fields and transported into surface waters is in part related to the particle size, charge and density, coupled with the velocity of the runoff stream. When particles mobilised by runoff encounter the relatively quiescent water column of a treatment wetland, much of the particulate matter can settle. The wetland vegetation is thought to contribute to particle removal through both sedimentation and filtration processes.

Particles removed in this fashion accrue in wetlands as sediments. A portion of the P associated with these sediments can be buried and permanently isolated from the water column. Depending on the nature of the particles, some P will likely be liberated from the settled particles as a result of desorption and/or decomposition processes. Sediment-water interface micro-environmental factors (e.g., electron acceptor availability, oxidation-reduction potential, and pH) and the chemical composition of the water and soil (e.g., sulphur, iron, calcium and aluminium contents) influence how effectively P in settled particles is retained in wetland sediments (Richardson, 1999).

Dissolved phosphorus

Particulate matter comprises only one component of the P in agricultural runoff and wastewaters. Dissolved chemical attributes of the soil are imparted to ADWs as they pass over or through the soil profile in farm fields and catchments. The complement of dissolved constituents in ADWs, including P containing compounds, is site-specific, a product of the catchment soil chemistry, topography, climate, and land management practices. For example, the P contained in inorganic fertilisers and even in manures can be exported in dissolved, bioavailable forms, particularly if application rates exceed crop requirements (Sharpley, 1999).

Dissolved P forms that enter a treatment wetland range from being quite labile to extremely recalcitrant. Soluble reactive P (SRP), operationally defined as the P fraction that passes through a 0.45 µm filter and is analytically detected by the molybdate blue method, is readily available for assimilation by bacteria, phytoplankton and macrophytes. Soluble reactive P can originate from fertilisers, or from decomposition of organic materials such as vegetation and manures. By contrast, dissolved organic compounds, which range from simple sugars to high molecular weight compounds, must first be broken down into more labile forms before their associated P can be assimilated by aquatic biota (Newman and Robinson, 1999). Some P is readily released from dissolved organic compounds following exposure to UV radiation or to enzymes (Wetzel *et al.*, 1995). In wetlands that successfully treat P to extremely low concentrations, recalcitrant DOP and particulate P compounds often comprise the bulk of the outflow P (Dierberg *et al.*, 2002a).

In wetlands, the labile dissolved P (the SRP) form is removed via biological uptake (by bacteria, phytoplankton, periphyton, and macrophytes), adsorption to chemical compounds (iron, aluminium and calcium) in soils and sediments, and chemical reactions in water columns (precipitation and/or co-precipitation). Phosphorus uptake rates and mass storages vary among ecological compartments, but regardless of the mechanism for SRP removal from the water column, the ultimate sink for P in wetlands is sediment.

The concentrations of constituents such as iron and aluminium in the water column may influence wetland P sequestration capabilities. In watersheds with little iron and aluminium, such as in the Florida Everglades, calcium can play a dominant role in P cycling. During photosynthesis, aquatic macrophytes can elevate surface water pH, thereby stimulating $CaCO_3$ precipitation. Precipitated calcium carbonate can provide sorption sites for dissolved phosphate ions (Dierberg *et al.*, 2002b). Thus, the influence of catchment water hardness on P retention in downstream treatment wetlands may be substantial.

Because wetlands generally are effective at removing particulate matter, it is possible that wetlands receiving P primarily in a particulate form exhibit higher mass P removal rates, at least on a short-term basis, than those dominated by dissolved P inputs. A complicating factor to such an assessment is that the cycling of P within wetlands is quite complex. Phosphorus can cycle among particulate, dissolved organic and dissolved inorganic compartments, on both a spatial and temporal basis, as water passes through a treatment wetland. For example, Dierberg *et al.* (in press) characterised the suspended algae entering and leaving a large 880 ha treatment wetland in South Florida and observed a change in algal density and speciation with passage through the wetland. Rather than simply "filtering" out the inflow algae, as evidenced by chlorophyll *a* reductions, the wetland was affecting species composition, which suggests cycling of P forms occurred between particulate and dissolved fractions.

Wetland phosphorus removal: design aspects

Several key parameters must be addressed in designing a treatment wetland for P removal. The most important of these are the expected flow rate (quantity and timing) of the ADW, the mean and range in ADW total P concentrations and the desired wetland outflow P concentration. Additional information that can facilitate treatment wetland design includes: climatic conditions at the site, since temperature can influence rates of microbial and macrophyte activity; the concentrations of P species (particulate vs. dissolved fractions) in the inflow ADW; and, the concentrations of other ADW constituents (e.g., nitrogen, oxygen demanding substances, total suspended solids, alkalinity, calcium, iron and aluminium contents).

Numerous treatment wetland design models are available, including steady-state and dynamic empirical models, many of which incorporate terms to address expected hydraulic characteristics of the treatment wetland (Kadlec and Knight, 1996). Mechanistic, process models also have been developed for treatment wetlands, but these invariably are too complex to have useful predictive capability. The simpler steady state models appear fairly effective at predicting either land requirements for a designated application, or the likely outflow P concentrations given a wetland parcel of known size. For predicting land area requirements, input parameters for such models typically include inflow P concentrations, outflow P concentrations, and a removal rate constant (K). Removal rate constant values were developed using historical performance data from other treatment wetland systems.

Another "rule of thumb" design approach is to evaluate previously developed relationships among mass P loading, outflow P concentrations, and mass P removal rates for treatment wetlands. Where historical data is available for wetlands that have operated in a similar climate and in a comparable inflow concentration range to the proposed treatment wetland, such relationships can be particularly relevant and useful for design. We compiled data from several sources to provide a range of potential scenarios encountered for wetlands used for removing P from ADWs around the world (Table 1). For each system, we identified mass P removal rates, inflow and outflow P concentrations, and hydraulic loading rates (HLR). To compare P removal among systems, we present performance parameters from these studies on a mean annual basis. An exception is the Braskerud (2002) study, where data was averaged across the entire period of operation (3-7 years).

Mass P loads for these ADW treatment wetlands vary widely, from about 1 g P m^{-2} yr^{-1} to over 100 g P m^{-2} yr^{-1} (Figure 1). Hydraulic loading rates also vary by about two orders of magnitude. These data suggest that mass P load is a better determinant of outflow P concentrations than HLR (Figure 1). Additionally, these data follow a commonly reported trend where wetland mass P removal rates increases with increasing P loads. While this is to be expected, since flow rate is a parameter embedded in both x- and y-axis terms, this relationship is useful in demonstrating that treatment wetlands operated at high mass P loads are capable of achieving quite high mass P removal rates (Figure 2).

With respect to design, the important factor to note is that interrelationships between mass P loading and outflow P concentrations dictate the amount of area required for the wetland. Wetland area requirements for P removal can be high, and they indeed are when compared to removal rates for other constituents. For example, using data reported for 15 treatment wetlands (see Kadlec, 2003), we calculate that the area required for the annual removal of 1 kg of P ranged from 1.4 to 1,378 m^2. The lowest area requirement for removing 1 kg P yr^{-1} was reported for systems receiving an extremely high mass P loading (1113 g P m^{-2} yr^{-1}). While mass P removal rates were impressively high, outflows from these systems contained high concentrations of P. By contrast, data from 18 systems described by Kadlec (2003) demonstrate

Table 1. Phosphorus loading and removal characteristics of wetlands used to treat agricultural runoff and wastewaters.

Location (source)	Treatment system	TP_{IN}* (mg P l⁻¹)	TP_{OUT}* (mg P l⁻¹)	TP load (g P m⁻² yr⁻¹)	Mass P removal g P m⁻² yr⁻¹	%	HLR (cm day⁻¹)
• New Zealand (Tanner and Sukias, 2003)	Surface and subsurface wetlands receiving waste stabilization pond effluent	16.7 - 42.4	9.6 - 35.6	173 - 2278	-4 - 296	-2 - 45	2.5 - 16.0
• Maryland, USA (Jordan et al., 2003)	Restored natural wetland receiving corn-soybean runoff	0.35 - 7.0‡	0.38 - 2.9‡	2.5 - 3.0	-0.28 - 1.8	-11 - 59	1.2 - 2.0
• Norway (Braskerud, 2002)	In-stream wetlands; fields in cereals with manure applied	0.17 - 0.43	0.10 - 0.27	91 - 191	18 - 71	21 - 44	66 - 181
• Finland (Koskiaho et al., 2003)	Boreal wetlands receiving farm runoff	0.073 - 0.57‡	0.063 - 0.22‡	0.93 - 11.1	-0.5 - 2.4	-6 - 62	1.9 - 14.3
• Florida, USA (SFWMD, 2004)	Wetlands receiving stormwater from irrigation/drainage	0.067 - 0.277	0.017 - 0.136	0.90 - 4.2	0.6 - 2.7	39 - 69	3.7 - 7.4

* Range in reported mean TP concentrations
‡ Calculated from reported mass removals, hydraulic loading rate and % removal

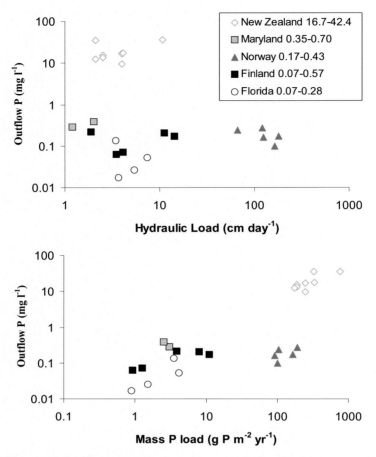

Figure 1. Outflow phosphorus (P) concentrations from treatment wetlands at five locations around the world, expressed as a function of (a) hydraulic loading rate and (b) mass P loading rate. The range of values given in the legend denotes annual mean ADW inflow TP concentrations (mg l⁻¹) for each respective wetland. Data sources for each wetland are provided in Table 1.

that the wetland area required for removing 1 kg of BOD_5 ranges from 0.5 to 70 m². Similarly, data from Kadlec and Knight (1996) demonstrate that an average of 76 m² of constructed wetland area is needed to remove 1 kg nitrate-N per year.

Case studies

We selected five ADW treatment wetland case studies from the literature, to illustrate performance under a range of inflow P concentration ranges and mass P loading rates.

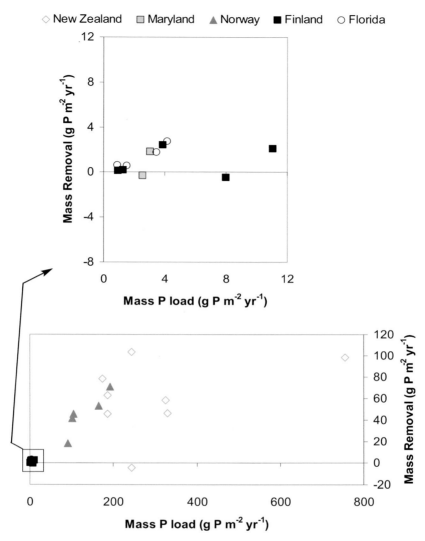

Figure 2. Mass P removal rate as a function of mass P loading for treatment wetlands at five locations. Panel on the top is presented at a finer scale to show differences among wetland systems operated at lower P mass loading rates. Data sources are presented in Table 1.

Florida, USA

Probably the most stringent application of wetlands for ADW P removal is the complex of Stormwater Treatment Areas (STAs) in South Florida. These are a group of six large treatment wetlands, totalling approximately 16,000 ha in area, that are being used to treat runoff from a 300,000 ha agricultural area prior discharge to the Everglades (SFWMD, 2004). The TP levels in runoff flowing into the wetlands typically range from 50 - 150 µg l⁻¹, and the target TP concentration is 10 µg l^{-1}. To date, several of the STAs have achieved outflow total P concentrations of 15-20 µg l^{-1} (SFWMD, 2004). At such low inflow concentration ranges, the resulting P loads are low (ca. 1- 3 g P m^{-2} yr^{-1}) as are the mass P removal rates (e.g., 0.8 - 1.3 g P m^{-2} yr^{-1}). Considerable wetland area therefore is needed, in this case ~1,000 m^2, to remove 1 kg of P annually.

New Zealand

At the other extreme, wetlands have been used to remove P from concentrated agricultural waste streams that contain much higher P concentrations. Tanner and Sukias (2003) describe wetland treatment of high P concentration (17 - 42 mg P l^{-1}) wastewaters from swine and dairy operations. Each wetland system was preceded by an anaerobic /facultative pond. Performance for three surface flow wetlands, two subsurface flow wetlands, and four surface-subsurface flow treatment wetlands is described by Tanner and Sukias (2003).

At moderately high HLRs of 2.5-16 cm d^{-1}, mass P removals exceeding 1000 g P m^{-2} yr^{-1} were achieved (Table 1). Such extraordinary load reductions were possible only under high P loading rates, which in this study resulted from both high P inflow concentrations and high hydraulic loading rates. While the mass P removal rates were high, it also should be noted that outflow P concentrations were an order of magnitude higher than wetlands at the other four locations (Table 1). This was a two-year study, so the long-term sustainability of wetlands receiving such high P loads is unknown.

Finland

Relationships between mass P loading rates and outflow P concentrations (Figure 1) suggest that increases in loading can lead to higher outflow P levels, and reductions in loading result in lower outflow TP concentrations. However, a potential problem with intermittent high P loads is that they may impair a wetlands ability to return to lower outflow P concentrations. In Finland, ADW was treated by three wetlands: the 0.6 ha Hovi demonstration wetland, the 0.48 ha Alastaro wetland, and a larger (60 ha), "semi-natural" Flyttrask wetland (Koskiaho *et al.*, 2003). The investigators monitored Hovi for one year, and Alastaro and Flyttrask for two years each. Agricultural fields covered 100%, 90% and 35% of the 12 ha, 90 ha and 2000 ha watersheds surrounding the Hovi, Alastaro, and Flyttrask wetlands, respectively. Snowmelt and heavy rains during spring caused extremely high flow rates and a pulse of dissolved nutrients in wetland inflows. Koskiaho *et al.* (2003) observed P export from the Alastaro wetland when high nutrient loads (11.1 g P m^{-2} yr^{-1}) in the first year were followed in a second year by 28% lower P loading (8.0 g P m^{-2} yr^{-1}). The larger Flyttrask wetland received lower mass loading (1.3 g P m^{-2} yr^{-1}) and showed no P release the following year, despite an equivalent percentage reduction in loading (to 0.93 g P m^{-2} yr^{-1}).

Norway

Four small wetlands (0.03-0.09 ha) were created in Norway by widening first-order streams located in agricultural watersheds that ranged from 22-148 ha in size (Braskerud, 2002). Flows were measured at v-notch weirs installed in the outflow dams, behind which the wetland surface water was retained. Three of four catchments contained 14-27% arable land (oats, barley) with the remainder forested, while one catchment contained 99% pasture land used for dairy cattle grazing. Streams flowed year-round, but summer HLR to the wetlands was ~50% of flows during the remainder of the year. Performance data was summarized for all years of operation, which ranged from 3-7 years among the four systems.

Because of the small wetland footprints relative to the watershed, HLRs were quite high (> 100 m yr^{-1}) (Table 1) (Braskerud, 2002). The range of inflow TP concentrations were similar to many ADWs (0.17 - 0.43 mg P l^{-1}), and while outflow concentrations were not much lower, the mass P removal rate was quite high (18-71 g P m^{-2} yr^{-1}) due to the extremely high HLR. Short retention time, coupled with low temperatures and biomass P demand in spring probably contributed to the low observed removal efficiencies (21 - 44 % P mass removed), relative to the other wetland systems considered (Table 1).

Maryland, USA

Jordan *et al.* (2003) presented two years of data from a restored natural wetland covering a contiguous 9% of its 14 ha watershed. Most (82%) of the watershed was cultivated (corn/soybean rotation), while the remainder was forested. Water entered the wetland from drainage leads, and exited the wetland through a standpipe outlet. Outflow P levels exceeded inflow P concentrations during the second year, presumably due to a two-fold higher HLR and lower inflow TP concentrations than the first year. These investigators suggest that a decrease in the inflow TP and PO$_4$-P concentrations in the second year may have caused release of P loosely sorbed to soil mineral components.

Enhancing long-term phosphorus removal performance

The case studies described above demonstrate several key trends related to P removal in treatment wetlands. First, wetlands that receive low P loadings, such as the 1- 2 g P m^{-2} yr^{-1} range of the South Florida STAs, are capable of achieving extremely low outflow P concentrations. Second, under variable P loading conditions, wetland outflow P concentrations do not necessarily respond (i.e., decline) immediately in response to load reductions. This phenomenon may be caused by some removal mechanisms (luxury uptake of P by biota and saturation of chemical adsorption sites during periods of high loadings) that do operate in proportion to mass loading rate. Finally, data from the New Zealand systems demonstrate that high mass P removal rates are attainable by treatment wetlands under high mass loading conditions. However, the sustainability of wetland P removal in these systems, particularly under prolonged, high P loads, is somewhat unknown.

Wetlands are thought to provide the most effective P removal performance during their first years of initial deployment. This is likely due to rapid vegetation growth and associated P assimilation upon initial flooding, coupled with ready availability of soil P sorption sites. However, as soil P sorption sites become saturated and vegetation reaches maximum standing crop levels, accrual of new sediment becomes the remaining pathway for long term P retention. Depending on the mass P loading rate and biogeochemical factors (e.g., renewal of sorption sites through inputs of calcium, iron and aluminium), declines in P removal performance may occur with time. Treatment wetlands that are highly loaded are the ones most likely to suffer from excessive sedimentation, which can not only impair P removal effectiveness, but also reduce overall water storage volume, particularly in the inflow regions of the system (Martinez and Wise, 2003).

Two techniques have been evaluated for enhancing sustainability of treatment wetland P removal. The first is periodic harvest of vegetation, and the second entails management of sediments. Wetland vegetation management for P removal has been assessed at various scales for at least three decades. The earliest work involved the use of the productive floating macrophyte, *Eichhornia crassipes* (water hyacinth) (Wolverton et al., 1976). Extremely high mass P removal rates have been achieved through harvest of water hyacinths and other productive aquatic macrophytes (exceeding 100 g P m^{-2} yr^{-1}) (Reddy and DeBusk, 1985). Similarly, work has been conducted for approximately a decade on using attached algae cultures ("algal turf scrubbers" or "periphyton filters") for P removal. Removal rates up to 0.73 g P m^{-2} d^{-1} have been reported for such systems (Craggs et al., 1996). While both attached algae and floating macrophytes such as *Eichhornia* and other species (e.g., *Lemna*) can provide effective P removal, this practice has never proven economically viable due to the high costs of plant harvesting, coupled with the low market value of the harvested biomass.

With respect to sediment management, it has long been recognised that wetlands receiving a high sediment load should be divided into at least two compartments, with a levee separating inflow from outflow regions. This is done to encourage the bulk of the allochthonous solids deposition to occur in the inflow region (Kadlec and Knight, 1996). There exists little information, however, on the effects of "front-end" sediment accumulation on wetland P removal performance, and whether or not pro-active management of such sediments can enhance P removal.

In 2003, our research team performed an *in situ* mesocosm study to evaluate the effects of sediment drydown and removal on overlying water column P concentrations of a Florida treatment wetland. This wetland was in operation for 16 years. The P content of the original mineral soil in the wetland was 48 ± 19 mg kg^{-1}, whereas the P content of accrued organic sediments near the inflow region was substantially higher, at 431 ± 170 mg kg^{-1} (Miner, 2001). We isolated portions of the water column and soil within 1.2 m diameter cylinders so that we could evaluate the effects of sediment drydown and organic sediment removal on P concentrations in the overlying water column. Measurements performed during 11 two-week batch incubations showed that sediment drydown alone provided a slight decrease in water column P levels, from 130 to 116 µg l^{-1}. By contrast, complete removal of the organic sediment layer sharply reduced water column TP levels, down to 62 µg l^{-1}. These data demonstrate that accrued wetland sediments indeed provide an internal contribution of P to the water column, and that their removal may contribute to improved performance. However, a problem with organic sediment removal is that it is expensive, and the technical effectiveness of this approach for full-scale wetlands has not yet been demonstrated. It is worth noting that removal of organic sediments from shallow lakes has had limited effectiveness in reducing water column TP concentrations (Moss et al., 1996; Ruley and Rusch, 2002). As an alternative approach to sediment removal in treatment wetlands, it may be possible to use chemical amendments to immobilise sediment P. Ann et al. (2000) reported that soils amended with calcium, iron and aluminium prior to wetland flooding reduced export of sediment P. Such an approach may prove less costly than bulk removal of accumulated sediments.

Conclusions

Phosphorus cycling within treatment wetlands is complex, with exchanges between dissolved and particulate P forms, and labile and refractory P forms, occurring dynamically on a spatial and temporal basis. Wetlands are capable of reducing P in ADWs to extremely low levels, but area requirements per unit mass of P removal can be extremely high. Unit area requirements appear to decline under higher mass P loading conditions, but this is achieved at the expense of higher outflow P concentrations. The gradual accumulation of P-enriched sediments with time can affect internal P cycling and limit long-term P removal effectiveness of treatment wetlands. Several techniques have been evaluated for improving wetland P removal effectiveness and sustainability, including routine vegetation harvest, removal of accumulated sediments, and chemical immobilisation of P in sediments. Such practices have been shown to work in pilot-scale systems, but their technical and economic feasibility for full-scale use remains to be demonstrated.

References

Ann, Y., K.R. Reddy and J.J. Delfino, 2000. Influence of chemical amendments on phosphorus immobilization in soils from a constructed wetland. Ecological Engineering **14** 157-167.

Braskerud, B.C., 2002. Factors affecting phosphorus retention in small constructed wetlands treating agricultural non-point source pollution. Ecological Engineering **19** 41-61.

Craggs, R.J., W.H. Adey, K.R. Jensen, M.S. St. John, F.B. Green and W.J. Oswald, 1996. Phosphorus removal from wastewater using an algal turf scrubber. Water Science and Technology **33** 191-198.

DeBusk, T.A. and W.F. DeBusk, 2000. Wetlands for water treatment. In: Applied Wetlands Science and Technology, edited by D.M. Kent, Lewis Publishers, Boca Raton, FL, 454 pp.

DeBusk, T.A., and F.E. Dierberg, 1999. Techniques for optimizing phosphorus removal in treatment wetlands, pp. 467-488. In: Phosphorus Biogeochemistry in Subtropical Ecosystems, edited by K.R. Reddy, G.A. O'Connor, and C.L.Schelske, CRC Press, Boca Raton, FL. 707 pp.

Dierberg, F.E., T.A. DeBusk, S.D. Jackson, M.J. Chimney, and K. Pietro, 2002a. Submerged aquatic vegetation-based treatment wetlands for removing phosphorus from agricultural runoff: Response to hydraulic and nutrient loading. Water Research **36** 1409-1422.

Dierberg, F.E., T.A. DeBusk, J. Potts and B. Gu, 2002b. Biological uptake vs. coprecipitation of soluble reactive phosphorus by "P-enriched" and "P-deficient" *Najas guadalupensis* in hard and soft waters. Verhandlungen Internationale Vereinigung fur Theoretische und Angewandte Limnologie **28** 1865-1870.

Dierberg, F.E., J. Potts and K. Kastovska, (in press). Alterations in the suspended algal abundance, distribution and diversity within a stormwater treatment area in south Florida, USA. Verh. Internat. Verein.Limnol.

Jordan, T.E., D.F. Whigham, K.H. Hofmockel, and M.A. Pittek, 2003. Nutrient and sediment removal by a restored wetland receiving agricultural runoff. Journal of Environmental Quality **32** 1534-1547.

Kadlec R.H., and R. Knight, 1996. Treatment Wetlands. Lewis Publishers, Boca Raton, FL.

Kadlec, R.H, 2003. Pond and wetland treatment. Water Science and Technology **48** 1-8.

Koskiaho, J., P. Eckholm, M. Raty, J. Riihimaki, and M. Puustinen, 2003. Retaining agricultural nutrients in constructed wetlands: experience under boreal conditions. Ecological Engineering **20** 89-103.

Martinez, C.J., and W.R. Wise, 2003. Hydraulic analysis of the Orlando Easterly Wetland. Journal of Environmental Engineering **129** 553-560.

Miner, C.L., 2001. Storage and partitioning of soil phosphorus in the Orlando Easterly Wetland treatment system. M.S. Thesis. Univ. Florida

Moss, B., J. Stansfield, K. Irvine, M. Perrow, and G. Phillips, 1996. Progressive restoration of a shallow lake: A 12-year experiment in isolation, sediment removal and biomanipulation. Journal of Applied Ecology **33** 71-86.

Newman, S., and J.S. Robinson, 1999. Forms of organic phosphorus in water, soils, and sediments. In: Phosphorus biogeochemistry in subtropical ecosystems, edited by K.R. Reddy, G.A. O'Connor, and C.L.Schelske, CRC Press, Boca Raton, FL. 707 pp.

Reddy, K.R., and Smith, 1987. Aquatic Plants for Water Treatment and Resource Recovery. Magnolia Publishing, Orlando, FL.

Reddy, K.R. and W.F. DeBusk, 1985. Nutrient removal potential of selected aquatic macrophytes. Journal of Environmental Quality **14** 459-462.

Richardson, C.J., 1999. The role of wetlands in storage, release and cycling of phosphorus on the landscape: A 25 year retrospective. In: Phosphorus biogeochemistry in subtropical ecosystems, edited by K.R. Reddy, G.A. O'Connor, and C.L.Schelske, CRC Press, Boca Raton, FL. 707 pp.

Ruley, J.E., and K.A. Rusch. 2002. An assessment of long-term post-restoration water quality trends in a shallow, subtropical, urban hypereutrophic lake. Ecological Engineering **19** 265-280.

SFWMD, 2004. Everglades Consolidated Report. South Florida Water Management District, West Palm Beach, FL, USA.

Sharpley, A.N. 1999. Global issues of phosphorus in terrestrial ecosystems. In: Phosphorus biogeochemistry in subtropical ecosystems, edited by K.R. Reddy, G.A. O'Connor, and C.L.Schelske, CRC Press, Boca Raton, FL. 707 pp.

Stuck, J.D., 1996. Particulate phosphorus transport in the water conveyance systems of the Everglades Agricultural Area. PhD dissertation. Univ. Florida.

Tanner, C.C., and J.P.S. Sukias, 2003. Linking pond and wetland treatment: performance of domestic and farm systems in New Zealand. Water Science and Technology **48** 331-339.

Wetzel, R.G., P.G. Hatcher, and T.S. Bianchi, 1995. Natural photolysis by UV irradiance of recalcitrant dissolved organic matter to simple substrates for rapid bacterial metabolism. Limnology and Oceanography. **40** 1369-1380.

Wolverton, B.C., R.M. Barlow, and R.C. McDonald, 1976. Application of vascular aquatic plants for pollution removal, energy and food production in a biological system. In: Biological control of water pollution, edited by J. Tourbier and R.W. Pierson, Jr., University of Pennsylvania Press.

The concept, design and performance of integrated constructed wetlands for the treatment of farmyard dirty water

R. Harrington[1], E.J. Dunne[2], P. Carroll[3], J. Keohane[4] and C. Ryder[5]
[1]National Parks and Wildlife, Department of Environment, Heritage and Local Government, The Quay, Waterford, Ireland
[2]Wetland Biogeochemistry Laboratory, Soil and Water Science Department, University of Florida/IFAS, 106 Newell Hall, PO Box 110510, Gainesville, FL 32611-0510
3Waterford County Council, Kilmeaden, Waterford, Ireland
4Geotechnical and Environmental Services, Innovation Centre, Carlow Institute of Technology, Carlow, Ireland
[5]Engineering Services, Office of Public Works, Dublin, Ely Place, Dublin 2, Ireland

Abstract

Integrated Constructed Wetlands (ICWs) are a holistic design specific approach to the use of constructed wetlands to improve water quality that has been adopted in Ireland. The explicit integration of water quality improvement such as the successful treatment of farmyard dirty water with landscape fit of ICW design, along with biodiversity enhancement, provides synergies that facilitate wetland system robustness and sustainability. Wetland influent characteristics such as ammonium concentration and volumes of incoming waters along with local site conditions, determine the scale and design of ICW systems. Performance data, in terms of phosphorus removal suggest that ICWs should be about twice the size of farmyard areas. However, this sizing is dependent on required effluent phosphorus concentrations. The improvement of water quality in the Annestown stream coincided with the construction and operation of farm ICWs within the Anne valley watershed. In general a more holistic approach to the management of waters within agricultural dominated watersheds is required.

Keywords: Integrated constructed wetlands, farmyard dirty water, landscape-fit, and ecosystems.

Introduction

The concept of "Integrated Constructed Wetlands" (ICWs) arose in the mid 1990s from the need to improve water quality as part of a wider community based ecological project in the Anne valley watershed in Co. Waterford, Ireland. The rural community of this 25 km² watershed had earlier identified the need for improved environmental infrastructure and had committed itself to improving the ecology and amenities of the valley's main water course, known as the "Annestown stream." The ICW approach and their use within a watershed have its origins in the "small watershed technique" and associated ecosystem studies developed by Bormann and Likens (1981) and Siccama et al. (1970).

Point sources of pollution were identified as being a major factor influencing the Annestown stream's water quality. After investigating various methodologies to prevent such pollution, an initial surface-flow constructed wetland to manage all water coming from one of the valley's farmyards was built and became operational in 1996. Fundamental to the wetland design was successful treatment of incoming waters and managing water flows; the need to fit the design into the landscape; and finally to enhance biological diversity of the site. These fundamental design principles were incorporated into all future constructed wetland projects in the valley. During the period 1999 to 2000, 12 ICWs for farmyards, which is about 75% of all farmyards in the watershed were built with full planning permission from the local planning authority. Financial and logistical support for these projects was provided by the European Union's Rural Development Fund (LEADER) with the central and local Government along with the individual farmers paying about one third of all costs. The construction of additional ICWs in the watershed is planned and it is expected that all farms will have an ICW system within the next few years. Numerous ICWs have been built in other parts of the county and country for the treatment of farmyard dirty water, sewage and industrial effluents.

The concept of integrated constructed wetlands

Constructed wetlands and ICWs specifically, are ecologically engineered systems rather than being just environmentally engineered to serve the single purpose of improving water quality (Mitsch and Jorgensen, 1989; Harrington and Ryder, 2002). Emergent vegetated wetlands, both natural and constructed, have innate abilities to cleanse water through physical, chemical and biological processes (Mitsch and Gooselink, 1993; Kadlec and Knight, 1996) and ICWs with their emphasis on ecosystem function in particular, are designed to mimic and harness these capacities. Integrated constructed wetlands have shallow water depths, are planted with emergent vegetation and use in situ soils in an effort to mimic the structure and processes found in natural wetland ecosystems. They are particularly focused upon achieving adequate hydraulic residence time especially during large precipitation events. They also strive to replicate a wider range of ecological conditions typically found in natural wetlands, which include those of soil, water, plant, and animal ecology. The use of local soil material and a wide variety of wetland plant species in ICWs are some of their features that distinguish them from a conventionally engineered "reed bed" system that typically uses a single species, the common reed *Phragmities australis* L. Furthermore, the explicit inclusion of "landscape fit", "biodiversity" and "habitat enhancement" into ICW design is fundamentally focused on providing additional values to the site, as well as water treatment. The larger land areas used in ICW design, compared with those used in other constructed wetland designs facilitates the incorporation of these environmental services. In addition, the larger area also provides for greater system robustness and sustainability particularly with regard to phosphorus retention.

The ICW concept also adopts an ecosystem-based approach to the management of farmyard waters, which is a more holistic approach to their management, rather than simply storage and disposal. It adheres to the "ecosystem approach" that has become the primary framework for action under the UNEP Convention on Biological Diversity (UNEP/CBD/SBSTTA/9/INF/4, 2003) and uses the ecosystem as the management unit. Ecosystems such as constructed wetlands are generally designed to be self-regulating systems, which negate the requirement for intense management; however this does not mean without management. The requirement of satisfying planning and regulatory authorities is achieved by demonstrating that the performance of an ICW does not, at least, impact negatively on the environment.

Farmyard dirty water

Agriculture in Ireland is recognised as one of the primary contributors to the eutrophication of Irish surface waters and the pollution of ground water (EPA, 2002). Agricultural inputs to surface waters can include both point and non-point source pollution. Point source pollution from agriculture is typically mismanaged farmyard dirty water. Farmyard dirty water can include yard washings, precipitation on open and roofed yard areas, dairy parlour washings, along with seepage of silage and farmyard manure effluents. The volumes of such waters are increased in most incidences by precipitation on impervious surfaces of farmyard areas.

At present, farmyard dirty water is managed in a variety of ways in Ireland. Currently, the most accepted method is that all farmyard dirty water generated is minimised. Relatively clean yard and roof waters are kept separate; the dirty water (typically yard and dairy washings) are collected and stored separately. These waters are then disposed of via land spreading and the relatively cleaner water is directed to drains and water courses (DAFF and DoE, 1996). This approach has both high capital costs and is labour intensive. The environmental impact of land-spreading is highly influenced by weather, soil type, topography and management (Richards, 1999; Bartley, 2003). Also, clean yard and roof waters, even if separated and maintained may be contaminated with suspended solids, sediments, avian faecal material, hydrocarbons and oxidised residues from roofing material. Thus, even the best conventionally managed farmyards may be polluting or have the potential to pollute. To ensure that that these actual or potential sources are contained a strategy of capturing all water flows from farmyards, their effective retention and on site treatment would seem to be a better management approach. In the economic, environmental, social and logistical context of present farming practices in Ireland, this need has particular relevance. In addition, if the Irish Government is to meet water quality criteria required under various European Union Directives, especially the Water Framework Directive, a more sustainable and coherent approach to managing all potential sources of pollution is required.

In general, the ICW approach strives to convert a conceived problem (the generation of farmyard dirty water) into a multi-dimensional resource by re-using or better managing the treated water, creating new ecological habitats within the agricultural landscape and using the area for education and recreation and as a general amenity to the rural community.

Integrated constructed wetland design

The ICW design requires that all existing polluted and potentially polluting water drain to the wetland being considered. The design also requires the effective containment of all contaminated water and that the wetland's water should not be a source risk of either point or diffuse pollution through inadequate wetland design and/or construction. Site assessment and design/planning guidelines for ICWs are dealt with in detail in these proceedings (Keohane *et al.*, 2005; Ryder *et al.*, 2005). Site assessment is an essential prerequisite to design. In summary it takes into account:

- Appropriateness of proposed sites.
- Soil type, geology, topography, coefficients of site uniformity.
- Site values for nature conservation and archaeology/built-heritage.
- Characteristics of influent (particularly ammonium concentrations) to determine particulate and dissolved components of incoming waters.
- An appropriate monitoring strategy, including consideration of adjacent wells, watercourses and ground water.
- The assimilative capacity of receiving water courses.

Wetland embankment height, in-flowing solids and accumulating detritus, determine the functional life-span of each segment of the ICW, with life-spans of 50-100 years expected on the basis of detrital accumulation and a minimum embankment height of 0.8 m. The potential visual aspect of the ICW design is important towards achieving empathy with both farm dwellers and the local community. Harmonious and curvi-linear shaping with level embankments aspires to a "natural" appearance of an ICW. The use of locally occurring wetland plant species is also the basis for establishing habitats and enhanced biodiversity appropriate to the locality that is consequently likely to further increase the robustness and sustainability of the system.

Determination of wetland area and influence of precipitation

Determination of wetland functional surface area for improving water quality has been generally calculated on the basis of first order rate decay and plug flow equations based on steady state conditions (Kadlec and Knight, 1996). Integrated constructed wetlands must treat water inflows of variable quality and quantity and are greatly influenced by precipitation. The ICW concept takes into account both extreme precipitation events as experienced in Ireland and the variable composition of influents in calculating wetland storage and sufficient treatment areas. Precipitation generated volumes generated in farmyard areas may be many orders of magnitude greater than that derived solely from dairy farmyard washings (Example 1).

Example 1

Volume of dairy washings
Typical farmyard dairy washings generated per cow = 50 l per day
100-dairy-cow herd = 5,000 l d^{-1} = 1825 m^3 yr^{-1}

Volume of precipitation on open farmyard areas
Approximate average yearly rainfall in Ireland = 1,000 mm yr^{-1}
Farmyard areas = 0.5 ha
Volume of yard runoff generated per year due to precipitation on open farmyard areas= 5000 m^3 yr^{-1}

Volume of precipitation on open farmyard areas during extreme event
50 mm during a 24 hr period
Open farmyard area = 0.5 ha
Volume of yard runoff generated during event = 250 m^3

Thus, using example one dairy washings are only 37% (on a yearly basis) of the volume of water generated from precipitation on open farmyard areas alone, whereas during a 24 hr extreme event, dairy washings can be as little as 2% of the volume of water generated from precipitation on open farmyard areas.

The cleansing effectiveness of surface flow wetland systems is typically based on having appropriate hydraulic residence times. In shallow-emergent-vegetated wetlands, such as ICWs, this depends on having sufficient functional wetland area with an appropriate length to width ratio and an emergent vegetation density. Surface hydraulic effectiveness of the ICW depends on:
- Segmentation of the wetland into a number of wetland cells of appropriate configuration.
- Avoidance of preferential flow.
- Dense vegetation stand.
- Managing water depth to ensure optimal functioning.

Infiltration of wetland waters from wetland bed and bank surfaces, along with evapo-transpiration have the capacity to create increased storage capacity (freeboard), thereby increasing hydraulic residence times. Such water loss buffers the impact of precipitation-generated fluxes through the ICW system.

Initial estimates for appropriate sizing of ICWs to manage incoming farmyard dirty water from the ICWs within the watershed was about 1.5 times the open farmyard or total interception area (open farmyard areas and roofed areas). This initial design estimate of wetland sizing was not always implemented by farmers in the construction of each of their ICW systems. This allowed the effectiveness of the ratio to be determined based on the results of our monitoring studies. We use this ICW sizing ratio both for effective storage of incoming waters, but also for the appropriate storage and retention of incoming phosphorus associated with farmyard dirty water. Average performance of six ICWs that were monitored for three years is shown in figure 1. Our results suggest that ICWs that were between three and four times the farm yard area, had outflow molybdate reactive phosphorus (MRP) concentrations of less than 0.5 mg l^{-1}. Toachieve an outflow concentration of 2 mg MRP l^{-1}, ICW to yard area ratio must be at lease two. We presently determine appropriate wetland area based upon having two times the interception area.

Another important design criteria essential for successful wetland performance is aspect ratio. Aspect ratio is defined as the average length of the wetland system divided by its average width. In our studies, aspect ratios ranged from 2 to 9, with a strong relationship between aspect ratio and effluent MRP concentration evident (Figure 1).

No significant relationship was found between ICW influent and effluent concentrations of MRP concentrations based on our monitoring studies (Figure 2); therefore a farmyard area based approach to determine sizing and anticipated performance appears appropriate in this circumstance.

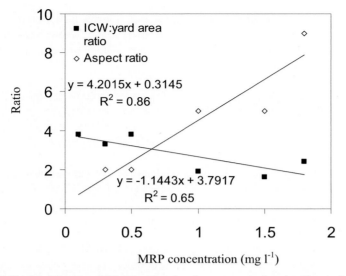

Figure 1. Relationship between molybdate reactive phosphorus concentrations in effluent of six integrated constructed wetlands and (i) wetland to yard area ratio, and (ii) average aspect ratio of the wetland systems.

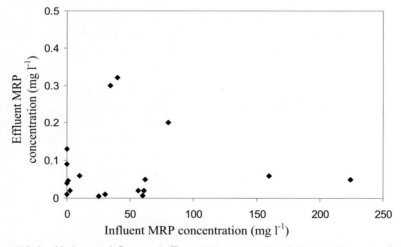

Figure 2. Relationship between influent and effluent MRP concentrations during the three years of monitoring between 2001 and 2004 for all 12 ICWs.

Limiting factors to wetland performance

During our monitoring studies, we observed no significant die-back in wetland vegetation for the 12 farm ICWs being monitored. Two other ICWs outside the valley that treated dilute pig slurry and meat processing wastewater showed vegetation stress and die-back at their respective points of in-flow. There seemed to be no relationship between vegetation die back

Nutrient management in agricultural watersheds: A wetlands solution

and seasonality; therefore other factors needed consideration. We hypothesised that high ammonia concentrations in incoming waters were responsible. Average ammonia concentrations of the dilute pig slurry were 340 mg l^{-1}, while average concentrations of the meat processing wastewater were 700 mg l^{-1}. A maximum ammonium threshold concentration of about 200-300 mg l^{-1} was observed in other ICWs that were used to treat meat industry wastewaters (Harrington, 2005). Therefore, incoming concentrations of ammonium may be a limiting factor in ICW design and performance.

Influents and effluent water quality

The performance of all ICWs in the Annestown stream watershed is presented in detail by Carroll *et al.* (2005). However, in general there was a large reduction in concentration of organic material suspended matter, nutrients and faecal bacteria, between ICW influents and effluents during the monitoring period between 2001 and 2004 (Table 1).

Effluent flows and impact on receiving waters

During periods of low precipitation (typically summer periods), ICW effluent flow tended to be correspondingly low for all ICWs and therefore there was no surface water discharge from ICW to receiving water (See Carroll *et al.*, 2005). This is a particularly important point, as during these periods receiving waters have typically low nutrient assimilation capacity.

It appears that the improvement of the Annestown stream since 1998 has coincided with the construction of the ICW farm systems (constructed between 1999 and 2000), which were treating about 75% of all farmyard dirty water generated within the watershed. The median MRP concentration for the stream that is the main receiving water body for all ICW surface water discharges was 0.02 mg L^{-1} for 2003. Biological water quality status of the stream has improved from an overall water quality rating of Q2 (seriously polluted) to a water quality rating of Q3/4 (slightly polluted) in 2001 (EPA 2002). Further evidence suggests that water

Table 1. Average influent and effluent water quality parameter concentrations of the 12 ICW systems monitored during 2001 and 2004.

Constituents	Farmyard dirty water ICW Influent	Surface water discharge ICW Effluent
COD[1] mg l^{-1}	2200	50
BOD[2] mg l^{-1}	1200	20
TSS[3] mm l^{-1}	700	20
Ammonium mg l^{-1}	80	0.5
Nitrate mg l^{-1}	<1	1.5
MRP mg l^{-1}	25	0.5
Total Coliformes cfu per 100 mls	220,000	<500

[1] COD = chemical oxygen demand.
[2] BOD_5 = five day biological oxygen demand.
[3] TSS = Total suspended solids.

quality has since improved to Q4 (unpolluted) (Mary Kelly-Quinn, personal communication, May, 2004).

ICW macroinvertebrate biodiversity

Although our results are preliminary, a study of representative ICWs within the watershed suggest that there is a wide range of macroinvertebrate organisms present in the latter segments of ICWs. One site-specific study showed that in upper wetland segments macroinvertebrate diversity was represented only by flies (Diptera); however in latter segments of the system macroinvertebrate diversity was represented by several orders that included Diptera, Hemiptera (true bugs), Ephemeroptera (mayflies) and Coleoptera (beetles) (Figure 3). These findings indicate that the addition of an ICW system within agricultural landscapes such as the Anne valley watershed provide suitable environments, primarily due to the presence of water, for local ecologies, such as macroinvertebrates.

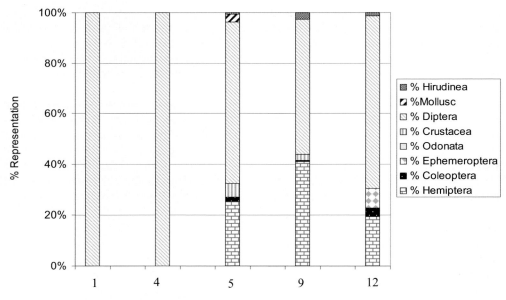

Figure 3. Percent representation of different orders of macro-invertebrate biodiversity with distance from ICW influent. Numbers on x-axis represent individual wetland cells in the ICW system at the Glanbia, Kilmeaden, cheese factory, Waterford.

Conclusions

The effectiveness of ICWs is largely dependent upon having appropriate wetland surface area to store and treat incoming waters such as farmyard dirty water. We observed the concentration of nutrients and contaminants in ICW influent had little influence on wetland effluent water quality and wetland area required; however incoming concentrations, especially ammonium concentration, was an important design criteria in terms of emergent wetland vegetation survival. Wetland vegetation are an important component of ICWs, as they provide the hydraulic resistance to surface flows and increase reactive surface area within ICWs.

The variation in ICW performance with regard to phosphorus removal was shown to be related to effective wetland area, which was determined relative to farmyard areas. In the pursuit of achieving low effluent phosphorus concentrations that are compatible with good water quality status in receiving waters, wetland areas should be twice the size of the farmyard interception area. This appears to be appropriate for ICW design and performance within the Anne valley watershed and these design approaches are being further researched. Future work on ICWs will consider among others, the type and content of influent including pesticide and hydrocarbon contaminants, wetland soil permeability, effects of weather and climate on wetland performance, vegetation dynamics, biodiversity within these wetland systems and the minimisation of surface water discharges to receiving waters.

We believe that the ICW design is a unique approach to the use of constructed wetlands to water quality management. They have the capacity to effectively treat contaminated water such as farmyard dirty water in a sustainable way. The management of such waters using ICWs can provide a more holistic approach to improving water quality within agricultural watersheds.

Acknowledgements

The authors wish to thank the participating farmers for their kind cooperation, Caolan Harrington for providing macroinvertebrate data. We thank Sinead Fox and Declan Halpin, Waterford County Council, for their contribution to this research.

References

Bartley, P., 2003. Nitrate responses in groundwaters under grassland dairy agriculture. Ph.D. thesis, Trinity College Dublin, Ireland.

Bormann, F.H., and G.E. Likens, 1981. Pattern and process in a forested ecosystem. Springer-Verlag, New York, 253 pp.

Carroll, P., R. Harrington, J. Keohane, and C. Ryder, 2005. Water treatment performance and environmental impact of integrated constructed wetlands in the Anne valley watershed, Ireland. In: Nutrient management in agricultural watersheds: A wetlands solution, edited by E.J. Dunne, K.R. Reddy and O.T. Carton. Wageningen Academic Publishers, The Netherlands.

Department Agriculture, Food and Forestry and Department of Environment, 1996. Code of good agricultural practice to protect waters from pollution from nitrates. DAFF and DoE, Dublin, Ireland.

EPA, 2002. Interim report: the biological survey of river quality 2001. Environmental Protection Agency, Johnston Castle, Wexford, Ireland.

Harrington, A., 2005. The relationship between plant vigour and ammonium concentrations in surface waters of constructed wetlands used to treat meat industry wastewaters in Ireland. In: Nutrient management in agricultural watersheds: A wetlands solution, edited by E.J. Dunne, K.R. Reddy and O.T. Carton. Wageningen Academic Publishers, The Netherlands.

Harrington, R., and C. Ryder. 2002. The use of integrated constructed wetlands in the management of farmyard runoff and waste water. In: Proceedings of the National Hydrology Seminar on Water Resource Management: Sustainable Supply and Demand. Tullamore, Offaly. 19th Nov. 2002. The Irish National Committees of the IHP and ICID, Ireland.Kadlec, R.H. and R.L. Knight. 1996. Treatment wetlands. Lewis Publishers, Boca Raton, 893 pp.

Keohane, J., P. Carroll, R. Harrington, and C. Ryder, 2005. Integrated constructed wetlands for farmyard dirty water treatment: a site suitability assessment. In: Nutrient management in agricultural watersheds: A wetlands solution, edited by E. J. Dunne, K.R. Reddy and O.T. Carton. Wageningen Academic Publishers, The Netherlands.

Mitsch, W.J., and J.G. Gosselink, 1993. Wetlands. 2nd Edition. Jon Wiley and Sons, Inc., New York, 722 pp.

Richards, K., 1999. Sources of nitrate leaching to groundwater in grasslands of Fermoy, Co. Cork. Ph.D. thesis, Trinity College Dublin, Ireland.

Ryder, C., P. Carroll, R. Harrington, and J. Keohane, 2005. Integrated constructed wetlands - regulatory policy and practical experience in an Irish planning context. In: Nutrient management in agricultural watersheds: A wetlands solution, edited by E. J. Dunne, K.R. Reddy and O.T. Carton. Wageningen Academic Publishers, The Netherlands.

Siccama, T.G., F.H. Bormann, and G.E. Likens, 1970. The Hubbard Brook ecosystem study: productivity, nutrients and phytosociology of the herbacious layer. Ecological Monographs **40** 389-402.

Integrated constructed wetlands: regulatory policy and practical experience in an Irish planning context

C. Ryder[1], P. Carroll[2], R. Harrington[3] and J. Keohane[4]
[1]Engineering Services, Office of Public Works, Dublin, Ely Place, Dublin 2, Ireland
[2]Waterford County Council, Adamstown, Co Waterford, Ireland
[3]National Parks and Wildlife, Department of Environment, Heritage and Local Government, The Quay, Waterford, Ireland
[4]Geotechnical and Environmental Services, Innovation Centre, Carlow Institute of Technology, Carlow, Ireland

Abstract

Constructed wetlands and integrated constructed wetlands (ICWs) in particular, are a relatively new phenomenon for Irish planning authorities. The authors have developed the concept of ICWs for the treatment of farmyard dirty water in particular. The background to Irish regulatory requirements is outlined, defining their general requirements for ICW systems. Experience with different local authorities is described with specific reference to the required supporting material needed for a proposed ICW site. Finally, the authors describe the proposal for a nationally agreed protocol for design and construction of ICWs to treat farmyard dirty water.

Keywords: planning, integrated constructed wetlands, farmyards, protocol, Ireland.

Introduction

Integrated constructed wetlands (ICWs) are a specific design approach to the use of constructed wetlands, which has been developed in Ireland. Fundamental to ICW design is successful treatment of incoming wastewater and positive benefits to the surrounding landscape. An ICW is a surface flow constructed wetland that somewhat mimics natural wetland processes such as the ability of wetland systems to cycle and retain incoming nutrients and contaminants. The ICW design also encompasses elements of landscape fit (Figures 1 and 2), biodiversity and habitat enhancement. In these facets it is unique and distinguishable from other constructed wetland systems. The theory and development of ICWs is explained in greater detail by Harrington and Ryder (2002).

The use of constructed wetlands such as ICWs to treat wastewaters is relatively new to Irish planning authorities. While various constructed wetland design formulae have been developed internationally, there has been little use of these in Ireland. The ICW design adopted a different approach and for this reason the authors faced a formidable task in demonstrating the value and efficacy of ICWs to planning authorities. However, during the past ten years, many ICWs have been constructed with full planning permission, which now serve as a strong resource base for further development of the concept.

Figure 1. Integrated constructed wetland (ICW) within the Irish landscape.

Figure 2. View of large integrated constructed wetland (ICW) in Annestown valley, Co. Waterford.

Irish legislative background

Several acts and regulations govern the planning and construction of ICWs. These include:
- Environmental Protection Agency Act, 1992
- Local Government (Planning and Development) Act 2000 (encompassing 1963 Planning Act and others)
- Local Government (Planning and Development) Regulations, 2001
- Local Government (Water Pollution) Act 1977 and (Amendment) Act, 1990
- Water Pollution Regulations of 1978, 1992, 1998 and 2003
- Protection of Groundwater Regulations, 1999.

Environmental Protection Agency Act 1992

This act established the role of the Irish Environmental Protection Agency (EPA) as the state agency responsible for environmental protection. It was allocated specific functions, which are fully outlined in Section 52 of the Act. These functions include the "licensing, regulation and control of activities for the purposes of environmental protection" as well as other monitoring, support, and liaison activities. Thus, it is the main agency governing matters relating to the environment and it is the main reference point for local planning authorities on matters relating to the environment. The EPA also establishes codes of practice and guidelines relating to environmental protection. Section 60 of this act clarifies the guidance role of the EPA in relation to the disposal of wastewater as follows; the EPA "may...specify...criteria and procedures...for the treatment or disposal of any sewage or other effluent to waters."

Local Government (Planning and Development) Act 2000

This planning act repealed all previous planning acts, and amalgamated all previous planning and development legislation under one umbrella act. Under section three, it defines development as 'the carrying out of any works on, in, over or under land," and while section four exempts "any land for the purpose of agriculture," section 32 clearly states that permission is required "in respect of any development of land, not being exempted development." It was left to subsequent regulations to clarify what was "exempted development."

Local Government (Planning and Development) Regulations 2001

These regulations defined those developments that were classed as "exempt." While class three, part three of schedule two exempted "the construction of any pond (which) shall not exceed 300 square meters," it is clear that, in general ICWs are not exempt from these planning and development regulations.

Local Government (Water Pollution) Act 1977 and (Amendment) Act 1990

This act defines quite clearly the obligation of citizens in relation to water pollution. Section three of the 1977 act states that "a person shall not cause or permit, any polluting matter to enter waters," while section four states, "a person shall not discharge or cause, or permit the discharge of any trade effluent or sewage effluent to any waters except under and in accordance with a license under this section." Agricultural effluent such as farm yard dirty water, as treated in an ICW, is classed as a trade effluent, and therefore its discharge to waters must be licensed.

Water pollution regulations

The 1978 regulations, first schedule, class one exempts "domestic sewage not exceeding in volume five cubic metres" from discharge licence requirements. This applies specifically to domestic sewage and does not include farm effluents such as farmyard dirty water.

The 1992 regulations section 42, state that a local authority such as a county council, "may specify conditions in a licence...in respect of a harmful substance specified in the second schedule." Phosphorus and ammonia, which are nutrients commonly found in high concentrations in farmyard dirty water are listed as harmful substances. Thus, conditions must be set for the licensing of their discharge. The 1998 water quality regulations outline, in the third schedule, various phosphorus quality levels for lakes and rivers, which are required.

The 1999 Protection of Groundwater Regulations list phosphorus and ammonia as "harmful substances." Section four states that "a Sanitary Authority shall not cause or permit the direct discharge to an aquifer save under and in accordance with a licence."

The 2003 European Communities (Water Policy) regulations sets out a policy for achieving "good status condition" for all waters in the context of the Water Framework Directive. These good status conditions in Ireland will be defined in 2005 by the EPA and are hoped to be achieved by 2015. These good status water quality criteria will consequently have a major effect on water quality within watersheds and this will require increasingly stringent water quality criteria of any system that is discharging to ground and/or surface waters.

Planning experience in Ireland

The recurring statistic in Irish water quality is that agriculture does contribute to the pollution of Irish waters (EPA, 2004). To quote from this year's EPA report "almost half the eutrophication of Irish rivers is due to agricultural sources, and over 70% of the phosphorus reaching inland waters, emanates from agricultural sources." Thus, the need for better farm management practices that both limits and mitigates nutrient loss from agriculture is obvious. The use of an ICW to treat farmyard dirty water is one possible management practice that maybe of use in reducing losses from agriculture.

Initially, the ICW concept was based on a demonstration project on a farm in the Anne Valley (Co. Waterford). It was through this demonstration project and subsequent monitoring data obtained that various design criteria were examined and the design further developed. This led to a large number of planning applications for ICW construction to Waterford County Council, the process of which is detailed below. Subsequently, a number of other local authorities, particularly in the Southeast of the country, have accepted planning applications and granted planning permissions to ICWs to treat farmyard dirty water.

Waterford County Council

In Waterford, the situation in relation to pollution from agricultural sources is no different from the rest of the country. The initial construction of a demonstration project on a local farm, the existence of a strong community organisation, and the critical presence of one of the authors of this paper (Rory Harrington) in the area, became a spur for the development of several ICWs within the Anne valley watershed. Through a series of community discussions, and site visits, a further 13 farmers initially agreed to construct an ICW on their lands, with the understanding that full planning permission would be required. All of these projects were supported in part by European and State funding. It was a cost share programme where farmers paid 50% of the construction costs. Full planning applications were prepared for these projects.

Concurrently with farmer discussions, a series of workshops were held by the authors, with Waterford County Council staff, to illustrate to them the ICW concept and outline previous monitoring results to demonstrate the efficacy of these systems. Further detailed discussions were held with planning staff to determine documentation requirements for ICW planning applications. Full planning applications were submitted for 13 farm ICWs. Following further requests for information by the County Council, all ICWs were granted planning permission with only minor conditions to construction, which included the requirement for site fencing, and the need for a discharge licences. Subsequently, the ICWs were constructed. Notable results within the Anne valley watershed since the construction of ICWs has been the steady improvement of the Annestown stream water quality, rising from a Q2 water quality status in 1999 to a higher water quality status value of Q3-4 value in 2004 (Harrington *et al.*, 2005).

Planning application requirements

In all of the planning applications submitted to the different county councils, there has been a consistency in relation to the documentation required. In general, the basic documentation required, as with any farm planning application includes:
- Completed planning application form. An ICW is considered as Class 3, which is the "The provision of structures for the purposes of agriculture." The planning application will also require details of existing farm operations and structures.
- Copy of advertisement from local and/or national papers detailing planning application.
- Copy of necessary site notice, required to be displayed prominently at the site during the application period.
- Payment of planning application fee.
- Site location and land ownership maps showing location(s) of site notice.
- Proposed site construction layout, and cross sections.
- Site characterisation report detailing trial hole and percolation test results.

Along with this basic documentation it is advisable to include a strong accompanying letter which, at the least, should give details on design and sizing of the ICW, and a broad description of the ICW concept and methodology. This document, as well as outlining the concept and operation of ICWs, also contains examples of typical expected effluent characteristics, for a properly designed, constructed and operated system. In a small number of cases, such as sites within an environmentally sensitive area, there has been a request by planning authorities for an Environmental Impact Statement (EIS).

Other planning authorities

While the major interest and subsequent construction of ICWs has been in county Waterford, there was considerable work done in other counties (Kerry, Cork, Limerick, Tipperary, and Wexford). The general acceptance of the ICW concept was slow. Planning authorities still find it difficult to assess the applications due to the lack of a "track record." There are a number of planning authorities that will not even consider an ICW planning application, on the grounds that it goes against the Irish Department of Agriculture and Food's guidelines for separation of farmyard dirty water, and because their performance cannot be "guaranteed" due to the lack of available data. It is proposed to address these issues that an agreed national protocol for ICW design and construction will be conducted.

Development of an ICW protocol

One of the particular requests from regulatory and planning authorities is a recognised guidance document or protocol, to enable an overall agreed consistent approach, based on present knowledge, developed for the design, planning, and construction of ICWs. This protocol is in the process of completion, and is due to be published shortly. The layout of the ICW protocol document will be:

Chapter 1: Introduction to the ICW concept and purpose of protocol.

Chapter 2: General environmental characterisation, identifying the critical environmental media at risk, in particular surface and ground waters, and the general suitability of an ICW for a proposed location.

Chapter 3: Site characterisation, detailing the required steps in carrying out a thorough site assessment and report.

Chapter 4: General methodologies for ICW design.

Chapter 5: Planning requirements for ICW construction and operation.

Chapter 6: Construction best practices, including health and safety protocols.

Chapter 7: Aftercare and site monitoring, detailing ongoing maintenance requirements and likely discharge licence requirements for monitoring purposes.

The protocol will be formulated such that it can accommodate new information and be reviewed with time.

References

Department of Environment Heritage and Local Government (DEHLG), 1978. Local Government (Water Pollution) Regulations, 1978 (S.I. No. 108 of 1978). Government Supplies Agency, Dublin.

DEHLG, 1992. Local Government (Water Pollution) Regulations, 1992 (S.I. No. 271 of 1992). Government Supplies Agency, Dublin.

DEHLG, 1998. Local Government (Water Pollution) Act, 1977 (Water Quality Standards for Phosphorus) Regulations, 1998 (S.I. No. 258 of 1998). Government Supplies Agency, Dublin.

DEHLG, 1999. Protection of Groundwater Regulations, 1999 (S.I. No. 41 of 1999). Government Supplies Agency, Dublin.

DEHLG, 2000. Planning and Development Act, 2000. Government Supplies Agency, Dublin.

DEHLG, 2001. Local Government (Planning and Development) Regulations 2001(S.I. No. 600 of 2001). Government Supplies Agency, Dublin.

DEHLG, 2003. European Communities (Water Policy) Regulations 2003 (S.I. No. 722 of 2003). Government Supplies Agency, Dublin.

DEHLG, 1992. Environmental Protection Agency Act, 1992 (S.I. No. 7 of 1992). Government Supplies Agency, Dublin.

Environmental Protection Agency (EPA), 2004. Ireland's Environment 2004 - the State of the Environment. EPA, Johnstown Castle, Wexford, Ireland.

Harrington, R, and C. Ryder, 2002. The Use of Integrated Constructed Wetlands (ICWs) in the Management of Farmyard Runoff and Waste Water, National Hydrology Seminar 2002: Water Resource and Management Sustainable Supply and Demand. Irish National Committees of IHP and ICID. The Institution of Engineers of Ireland. pp. 55-63

Harrington, R., E.J. Dunne, P. Carroll, J. Keohane, and C. Ryder, 2005. The concept, design and performance of integrated constructed wetlands for the treatment of farmyard dirty water. In: Nutrient management in agricultural watersheds: A wetlands solution, edited by E.J. Dunne, K.R. Reddy and O.T. Carton. Wageningen Academic Publishers, The Netherlands.

Integrated constructed wetlands for farmyard dirty water treatment: a site suitability assessment

J. Keohane[1], P. Carroll[2], R. Harrington[3] and C. Ryder[4]
[1]Geotechnical and Environmental Services, Innovation Centre, Carlow Institute of Technology, Carlow, Ireland
[2]Waterford County Council, Adamstown, Co Waterford, Ireland
[3]National Parks and Wildlife, Department of Environment, Heritage and Local Government, The Quay, Waterford, Ireland
[4]Engineering Services, Office of Public Works, Dublin, Ely Place, Dublin 2, Ireland

Abstract

Constructed Wetlands and Integrated Constructed Wetlands (ICWs) in particular are a relatively new phenomenon in Ireland. This paper describes an approach to site suitability assessment for the construction of an ICW. This will form part of a nationally agreed protocol on the design and construction of ICWs for the treatment of farmyard dirty water. The site assessment approach is based on a combination of desk study and on-site works. It uses a risk based format to address the potential impact on environmental receptors such as groundwater and surface water, flora, fauna, landscape and material assets.

Keywords: integrated constructed wetlands, site assessment, risk, surface water, groundwater.

Introduction

Integrated constructed wetlands are a specific constructed wetland design approach to the management of farmyard dirty water in Ireland. The decision to use an ICW, as opposed to any other management method, will be made on an evaluation of technological, environmental, economic, and logistical criteria, along with personal preference of the farmer. A site suitability assessment is of key importance in this decision making process. It is important that a systematic and logical approach is followed to allow the suitability of the site to be assessed as early as possible in the decision making process, so that resources are used effectively. It is likely that any development of an ICW to treat soiled water such as farmyard dirty water will require planning permission. The relevant planning authority will need to be provided with adequate information in a standard, easily understood and logical format in order to assess the proposed ICW development.

This paper presents an approach to assessing site suitability with the objective of collecting sufficient information to (i) determine if an ICW can be developed on a particular site, (ii) demonstrate what impact the construction of an ICW may or may not have on the surrounding environment, and (iii) provide adequate information for an appropriate site-specific wetland design. Site characterisation as a component of the assessment procedure includes a desk study, a visual assessment, and on-site tests to determine site suitability.

Site assessment in the context of the ICW approach

The ICW approach adopts a holistic "ecosystems approach" to the management of water resources. The objective is to use a constructed wetland that is designed on a site-specific basis to achieve low volumetric discharge along with discharge waters that have good water quality. The overall wetland design should be integrated into the landscape and should provide an additional ecosystem, as well as having additional benefits for the landowner and the countryside in general. Whilst ICWs have been used for a diverse range of applications, this paper deals solely with their use on Irish farms to treat soiled or farmyard dirty water. Farmyard dirty water is generated by rainfall runoff from open yards dairy washings, along with silage and manure seepage effluents.

Site characterisation for a potential ICW should have several objectives. These are shown in Table 1.

Table 1. Components of site characterisation for an integrated constructed wetland (ICW).

Objectives	Implication for site characterisation
Integrate design of wetland into the surrounding environment	Ensuring that sufficient information is gathered to enable the risk to be controlled by natural protection supplemented only where necessary with hard engineering measures.
Optimise the natural treatment processes	– Ensure that the nature and properties of the wetland influents are known – That adequate land space is available – Minimise energy inputs
Obtain good landscape fit	Map site topography
Enhance biodiversity of local ecology	Achieving a comprehensive understanding of the local ecological setting so that a diverse range of wetland plants and open water areas can be incorporated into the wetland design

A risk based approach

A risk based approach to the management of water resources is common among scientific disciplines, regulatory agencies and the private sector, as the environment is too heterogeneous to ensure 100% protection at all times. This means that we need to rely on risk minimisation as a first principle. The concept of risk is therefore important in the overall approach to site suitability assessment and design of an ICW. A risk based assessment provides a framework for evaluating and managing pressures and impacts on identified environmental receptors. We propose to use the hazard-pathway-receptor model. Risk can be defined as the likelihood or expected frequency of a specified adverse consequence. Applied to ICWs, it expresses the likelihood of damage or contamination arising from the construction or operation of an ICW (Hazard). A hazard presents a risk when it is likely to affect something of value such as surface and/or groundwater (Receptor). An impact can only occur if a linkage (Pathway) is established between the hazard and the receptor. Protection, like risk, is a relative concept and there is an implied degree of protection, rather than absolute protection. An increasing level of protection is equivalent to reducing the risk of damage to the receptor. Moreover, choosing the appropriate level of protection necessarily involves placing a relative value on the protected entity.

Key environmental receptors

In the context of an ICW the key environmental receptors are outlined in Table 2.

Table 2. Environmental receptors for an integrated constructed wetland (ICW).

Receptor	Issues
Surface water	Generally, the ICW will discharge to a surface water feature; therefore a discharge licence will be required. The receiving water may have a water quality rating based on its existing chemical and biological status.
Groundwater	Unless the wetland is artificially sealed with a synthetic liner, there will be some percolation to the aquifer, which will have a resource value in keeping with national groundwater protection criteria.
Topsoil and subsoil	Site topsoil will be used as rooting media for wetland plants. Suitable subsoils will be used to line wetland surfaces. With time topsoils will accumulate nutrients, which will, at some stage, have to be removed and/or reused.
Landscape	An appropriate landscape fit is a key objective of the ICW concept, it is therefore important that the wetland is designed cohesive to the topography of the landscape.
Flora and fauna	Construction of the wetland will impact flora and fauna in the environs of a particular site. The wetland itself should enhance the biodiversity, but care will need to be taken to ensure that legislatively protected areas such as Special Areas of Conservation (SACs) or Natural Heritage Areas (NHAs), which are legislatively protected, are not negatively impacted.
Air	Minor odours, can be associated with treatment wetland systems, and their impact on air quality should be assessed.
Cultural heritage	A desk study should identify any known heritage sites of importance. Care during construction must be exercised to eliminate damage to possible undiscovered sites.
Human	The potential impact on the farm enterprise and neighbouring landowners and properties should be assessed.

Hazard characterisation

The principal contaminants, which constitute the hazard, are related to quality of the farmyard dirty water. Components of dirty water can include animal wastes (faeces and urine), feedstock, dairy wash water, slurry and silage seepage, and farmyard runoff. The dirty water produced in farmyards is related to precipitation. Thus, the volumes of dirty water generated in farmyards can be very variable. The volumes of farmyard dirty water generated is generally are calculated from precipitation on open yard areas. For example, 1000 mm yr^{-1} of precipitation on an open yard area of 5,000 m^2 can generate up to 5,000 m^3 of dirty water per annum. In a storm event, where 100 mm of precipitation can fall during a two day period, the volume of dirty water produced for the same contributing area could be 250 m^3. This example highlights that there can be large fluxes to the ICW and thus, the wetland should be sized to accommodate these extreme events, without compromising dirty water treatment performance.

Chemical characteristics of farmyard dirty water can also be variable (Table 3). However, it is possible to provide a list of the typical parameters that constitute farmyard dirty water and to provide a range of concentrations for these parameters.

Post wetland treatment the effluent will generally enter a nearby surface water feature such as a stream or river; therefore the surface water body will be the potential receptor. Furthermore, because there will be some percolation from the wetland surfaces, the underlying soil and nearby groundwater body will also be a receptor. Typical characteristics of ICW discharges to surface (Table 4) and ground waters (Table 5) are shown. The volumes of treated waters discharging from the wetland to surface waters are variable and there can be long periods of time, particularly during summer months, when no discharge from the wetland occurs.

Table 3. Concentrations of some water quality parameters that are typically measured in Irish farmyard dirty water. Values (mean, range and standard deviation) are based on three years of monitoring data from farms in Co. Waterford that use ICWs to treat farmyard dirty water. (Source: Carroll et al., 2005).

Constituents	Farmyard dirty water (mean) (mg l^{-1})	Range (mg l^{-1})	Standard deviation (mg l^{-1})
COD[1]	2200	20 - 90,000	8,000
BOD$_5$[2]	1200	2 - 60,000	5,000
Ammonium (NH$_4^{+}$ -N)	80	0.1 - 1900	170
Nitrate (NO$_3^{2-}$ N)	<1	<1 - 10	2.5
Phosphate (PO$_4^{3-}$ P)	25	0.01 - 900	70

[1] COD is chemical oxygen demand.
[2] BOD$_5$ is five day biological oxygen demand.

Table 4. Characteristics of ICW treated waters that were discharged to surface waters. Values (mean, range and standard deviation) are based on three years of monitoring data from farms in Co. Waterford that use ICWs to treat farmyard dirty water. (Source: Carroll et al., 2005).

Parameter	Wetland treated waters (Mean) mg l^{-1}	Range mg l^{-1}	Standard deviation mg l^{-1}
COD	50	5 - 280	34
BOD	20	<1 - 180	20
TSS[†]	20	<1 - 190	25
Ammonium (NH$_4^{+}$ -N)	0.5	0.1 - 17	2.0
Nitrate (NO$_3^{2-}$ N)	<1	<1 - 17	3.0
Total phosphorus	1.0		
Phosphate (PO$_4^{3-}$ P)	0.5	0.01 - 7	0.8
Faecal coliforms per 100 mls	<500	<100 - 9,000	1,320

Table 5. Characteristics of ICW treated waters that were discharged to ground waters. Values (mean, range and standard deviation) are based on three years of piezometer monitoring data from farms in Co. Waterford that use ICWs to treat farmyard dirty water. (Source: Carroll et al., 2005).

Parameter	Wetland treated waters (Mean) mg l^{-1}	Range mg l^{-1}	Standard deviation mg l^{-1}
Ammonium (NH_4^+ N)	4	0.3 - 17	4
Nitrate (NO_3^{2-} N)	0.2	<0.2 - 9	1.5
Faecal coliforms per 100 mls	25	<1 - 50	27
Phosphate (PO_4^{3-} P)	< 0.01	<0.01 - 0.07	0.04

In terms of percolation waters from wetland bed surfaces, the wetland bed should have a permeability of 1×10^{-8} ms^{-1}. The percolation rate through this base will be approximately 0.9 mm d^{-1}. For an ICW with a bed area of 10,000 m^2, this equates to 9 m^3 d^{-1} of percolating waters leaving the wetland.

Site suitability

There are a number of pre-requisites, which must be satisfied before embarking on an assessment of site suitability for an ICW. These include set back distances from an ICW to potential receptors. Set back distances may vary from time to time and will depend on planning authority. The following are proposed for ICW sites:
- A minimum distance of 60 meters from any well or groundwater spring used for potable water.
- A minimum distance of 300 meters up gradient of a public water supply borehole, where an inner protection area has not been identified.

Geographic locations where ICWS should not be built include:
- Sites within the inner protection areas of a water source, where the vulnerability rating is classified as extreme.
- Within 25 meters of a house dwelling.
- Within the area of a mature tree root systems.
- Sites that are underlain by karst limestone features, where the possibility of collapse cannot be ruled out.
- Sites where construction of the ICW will damage or destroy a site of natural or cultural heritage.
- Sites where sensitive surface waters will receive wetland discharge.
- Sites where adequate land area is not available for the construction of an ICW.

Surfacewater protection requirements

Discharge rates from ICWs are variable and often there are long time periods, particularly during summer months, when no discharge occurs. To assess the assimilative capacity of receiving surface waters, the use of average annual flows of these waters should be used to determine whether there is sufficient flow in the receiving stream or river to assimilate wetland discharged waters. Information on flows of water courses for a particular location can generally be obtained from the national or regional hydrometric monitoring agency such as the Irish Environmental Protection Agency.

Groundwater protection requirements

One of the objectives of the desk study is to gather hydrogeological information such as aquifer classification and groundwater vulnerability for a specific site. The purpose of the site works is to confirm or modify this setting such that there is minimum risk to groundwaters. There are a number of basic requirements which provide adequate protection of groundwater. In general, all ICWs should be underlain by at least one meter of moderate or low permeability subsoil, with the upper 0.5 m enhanced where necessary, to achieve a permeability of 1×10^{-8}m s^{-1}. The definition of low permeability material is based on a description from trial hole tests, together with a "T" value of 30 or greater from the percolation test undertaken during site characterisation. There will be additional requirements, where the potential risk is higher. These include:

- Where a regionally important aquifer is present, the total thickness of subsoil underlying wetland surfaces should be increased to 1.5 m, with the upper 0.75 m enhanced if necessary, to achieve a permeability of 1×10^{-8}m s^{-1}.
- Where high permeability sand and gravel is encountered in subsoils and is in hydraulic continuity with the water table, the ICW can only be constructed if 1m of low permeability material is laid over the sand and gravel, with the upper 0.75m enhanced, if necessary, to achieve a permeability of 1×10^{-8}m s^{-1}.
- Post-construction, site monitoring should be used to assess water quality in percolating water from wetland bed surfaces. This could be done using in situ structures such as lysimeters and/or piezometers.

The competency of the assessor

The person undertaking the site assessment must have:
- An appropriate recognised qualification.
- Competency to collect desk study information and interpret its findings.
- Carry out the visual assessment and interpret the findings.
- Capable of identifying situations where a specialist needs to be engaged in site assessment to assess the likely impact on a particular receptor.

Site characterisation

The following key steps are proposed for appropriate site characterisation during site suitability assessment:
1. Desk study and collation of information.
   a. Preliminary consultation.
   b. Collation of relevant information.
   c. Desk study assessment.
2. Visual Assessment.
   a. Characterisation of hazard - farmyard inventory.

   b. Evaluation of receptor sensitivity.
3. Site tests:
   a. Trial hole.
   b. Percolation test.
4. Decision process and preparation of recommendations.

Desk study and collation of information: preliminary consultation

A preliminary consultation with the project proposer/farmer is strongly advised. The purpose of the preliminary consultation is to (i) establish motivating factors for the farmer wishing to build an ICW, (ii) establish the current farmyard dirty water management practice, (iii) establish in general terms, what will be entering the wetland, and (iv) provide the farmer with some general facts on ICWs. The farmer may have one or more reasons for considering the use of an ICW, and these may include ongoing fines or difficulties with a local authority, state agency or neighbours, requiring improvements in environmental performance, or it maybe a desire on the part of the farmer to improve the environmental performance of their farming practices. Information collected during consultation should be included in the site assessment. This consultation may help to identify potential receptors at risk.

The current on site management regime for farmyard dirty water should be recorded. Is the farmyard dirty water not managed at all, or is it managed using conventional land spreading practices? It may be possible to improve present management practices rather than constructing a wetland for the treatment of farmyard dirty water. Present farmyard dirty water management needs to be discussed with the farmer and/or the farmer's agricultural adviser.

The characterisation of farmyard dirty water that will enter the wetland needs to be established in broad terms at this stage, as will the size of open farmyard areas, which can contribute to the generation of farmyard dirty water. An approximate wetland design and size can be discussed with the farmer and/or farmer's agricultural adviser in order to give them an estimate of the required land area for ICW construction.

Collation of relevant environmental data during the desk study is also important. The purpose of this is to:
• Obtain information relevant to the site.
• Identify targets at risk.
• Establish if there are site restrictions.
• Shortlist site location options, if more than one location is being considered.

Information and site data that should be collated at this stage are outlined in Table 6.

Desk study assessment

The information collected from the desk study should be examined. Having reviewed the topographical maps surface water features, possible topographical constraints, and the presence of any mapped areas of heritage should be identified. This may lead to the discovery of further information related to surface water and heritage from the relevant sources. To avoid accidental

Table 6. Site information required for appropriate site characterisation.

Media	Sources of information
Topography	Maps, Ordnance Survey maps
Surface Water	Environmental Protection Agency, Office of Public Works, Local Authority
Geological and hydrogeological	Geological Survey of Ireland
Natural and cultural heritage	Local Authority, Department of Environment, Heritage and Local Government
Climate	Met Eireann
Site soils and drainage	Farmer/advisor
Public utilities	Utility providers
General planning	Local authority

damage, a trial hole assessment or percolation test should not be undertaken in areas which are at, or adjacent to sites of national legislative protection such as SAC's and NHA's.

The geological information collated should indicate the potential of encountering karst or high resource value aquifers during site assessment. Site soil information will have highlighted the possibility of encountering gravel or potentially low permeability material on site. Once the aquifer and vulnerability classes are established reference to groundwater protection matrix will determine the appropriate response and requirements for of that response. The on-site assessment will later confirm or modify such responses.

The recorded climate data for the region will have been collected, and any constraints relating to past and present land drainage systems, utilities, and planning will be identified. Any information on satisfactory or unsatisfactory local experience with ICW's can be incorporated at this time to complete the desk study assessment. By this stage it may be possible to eliminate sites that present too many constraints where construction of an ICW would not be appropriate.

Visual assessment

The purpose of the visual assessment is to:
- Verify desk study findings.
- Make an on-site assessment of hazards.
- Evaluate the sensitivity of the identified receptors (Table 7), in particular receiving watercourses and groundwater features.
- Select site for ICW construction.

Characterisation of hazard

The site work should begin with a visual assessment of farmyard management and operations, such that present farm practices are evaluated. The time of this characterisation is important, as problems with the management of farmyard dirty water may not be apparent on a dry summers day; however problems maybe extreme during wet winter months. It is useful to make a sketch of the farm layout, or use a map already obtained from the farmer or agricultural adviser to evaluate the farmyard. The type of farming practice is important for characterisation of hazard. Information such as area devoted to different land management practices, approximate land area devoted to each practice, number of animals and the duration these animals spend outside during the year should be determined. Most, if not all information should be readily available from the farmer and/or their agricultural adviser.

The quality and quantity of farmyard dirty along with its sources should be evaluated. This will require measuring contributing open yard areas. Annual quantity of dirty water produced per contributing area should be determined. For example runoff generated from open yard areas can be determined by knowing annual rainfall and open yard areas. Farmyard roof waters may contribute to the overall hydraulic loading on to the ICW. The construction and layout of slurry pits, dung heaps and silage clamps need to be evaluated to determine whether they contribute to run-off from farmyard areas. The general management of dairy washwaters (if present) should also be evaluated. This will entail an assessment of: quantity of waters used, number of washes per day, collection, suitable storage and treatment of those waters.

Photographs should be taken on site to help characterise potential hazards, visually record the general farmyard and farm layout, type of building structures, and to record the various features of interest on site.

It is important that the farmer is present and participates during this initial on site characterisation process to discuss the various changes in farm activities that may need undertaking prior ICW construction.

The relevant set back distances from the ICW site should be referenced and adjustments made as necessary. Sites that do not satisfy these requirements may be identified at this time. The nature of the hazard should be fully understood at this stage and this will need to be verified following a topographical survey of the site. The proposed receiving water (receptor) will be fully assessed and water quality samples may have to be taken for baseline characterisation. A good indication of the soils, geological and hydrogeological setting will be established, and this will be verified as part of the trial hole and percolation test, which will be discussed later.

Table 7. Identified receptors and their key issues regarding their on site assessment.

Media	Key issues
Topography	Scope of topographical survey, slopes, landuse, landscape fit
Surface water	Density of surface water features, visual assessment of receiving water
Public utilities	Verify desk study findings
Cultural heritage	Verify desk study findings
Human	Distances from human receptors
Flora and fauna	Verify desk study findings, examine any existing wetland vegetation in nearby environs
Land drainage	Verify desk study findings
Groundwater	Identify wells, determine groundwater levels
Climate	Prevailing wind direction on site
Soils	Examine any on site or nearby open excavations

Site construction constraints imposed by the presence of natural or cultural heritage features will be assessed, and the potential of encountering land drainage systems will need evaluating.

Site tests

Trial hole

The trial hole tests are probably the most critical element of the on site characterisation. The purposes of the trial hole are to determine (1) the depth to the water table, (2) the depth to bedrock and (3) the soil and subsoil characteristics of the site. Trial hole digging should be arranged to coincide with the visual assessment. A minimum of three trial holes at the site should be dug to at least 2 m below the proposed base level of the ICW. Appropriate subsoil characteristics for ICW construction are shown in Table 8.

Percolation test

A percolation test typically assesses the permeability of a subsoil. In practice this test measures the time it takes water levels to fall a set distance within a previously dug percolation hole. The actual procedures for carrying out a percolation test are outlined by the Irish EPA's 2000 manual on "Treatment Systems for Single Houses." A minimum of four percolation holes should be dug on site. Similar to the trial hole tests, the number of percolation holes should be increased where conditions are found to be variable. The results of the percolation tests are expressed as a "T value." This is the average time in minutes for the water level to fall 25 mm within the percolation hole. For the percolation test, the requirement for a low permeability base suggests that the drop in water level should be in the region of 1 mm or not measurable after a 24 hour period. Where this is not the case, then some form of soil enhancement works will be required on subsoils to create conditions where water can be retained to the specified level. In general, it is considered that a "T" value greater than 30 with mainly cohesive material will indicate conditions that can be enhanced.

Table 8. Subsoil characteristics which indicate satisfactory sits soil characteristics required for construction of an ICW.

Factors	Characteristics
Colour	• Free draining unsaturated soils/subsoils are in an oxidised state; therefore they should have a brown, reddish brown and/or yellowish brown colours. • Saturated soils/subsoils are in a reduced state and should exhibit dull grey or mottled colours.
Structure and texture	Subsoils should comprise of high percentages of clay and/or silt. The presence of sands and gravels may present site construction difficulties.
Depth to rock	There must be sufficient subsoil overlying bedrock for the protection of groundwater aquifers.
Depth to water table	Unlikely to be a problem in low permeability subsoils, but can be a problem in highly permeable subsoils.
Assessment of soil permeability	Sites underlain by soils with low permeability material may not require enhancement to satisfy planning requirements.

Decision process and preparation of recommendations

After undertaking the different components of the site suitability assessment the information collected is used to determine whether an ICW can be constructed on any particular site. In conclusion, table 9 summarises the main components of the desk study and the on site assessment, which are the two most important components in determining the suitability of a site for the construction of an ICW.

Finally, at the end of the site suitability assessment, one should be able to: determine if the ICW can be developed on a particular site; provide adequate information to enable a site-specific ICW design; and demonstrate that the construction of an ICW is the most appropriate management option for that particular site.

Table 9. Summary of information collected from desk study and overall site assessment process.

Information collected	Key issues	Implications
Topography	Slopes and land profile	Design, overall site layout, and potential landscape fit of ICW
Surface water	• Receptor sensitivity • Receiving water capability • Possible flood levels	• Set back distances • Discharge licence • Wetland design for extreme events
Hydrogeological setting	Receptor sensitivity	• Design of ICW base • Set back distances • Groundwater monitoring
Depth to rock	Pathway assessment	Design, base details, site suitability
Subsoil type	Pathway assessment	Design, construction site suitability
"t" value	Pathway assessment,	Design, construction site suitability
Farm inventory	Hazard assessment	Design and sizing of ICW
Cultural heritage	Distances from sensitive sites	Statutory and planning.
Natural Heritage	Distances from sensitive sites	Statutory and planning
Climate	• Rainfall • Receptors down wind	• Design and sizing of ICW • Orientation and layout of ICW
Human	Possible odours	Set back distances
Housing	Proximity	Set back distance

References

Environmental Protection Agency. 2000. Wastewater treatment manuals: treatment systems for single houses. EPA, Wexford, Ireland.

Carroll, P., R. Harrington, J. Keohane, and C. Ryder, 2005. Water treatment performance and environmental impact of integrated constructed wetlands in the Anne valley watershed, Ireland. In: Nutrient management in agricultural watersheds: A wetlands solution, edited by E.J. Dunne, K.R. Reddy and O.T. Carton. Wageningen Academic Publishers, The Netherlands.

Nutrient management in agricultural watersheds: A wetlands solution

Water treatment performance and environmental impact of integrated constructed wetlands in the Anne valley watershed, Ireland

P. Carroll[1], R. Harrington[2], J. Keohane[3] and C. Ryder[4]
[1]Waterford County Council, Kilmeaden, Waterford, Ireland
[2]National Parks and Wildlife, Department of Environment, Heritage and Local Government, The Quay, Waterford, Ireland
[3]Geotechnical and Environmental Services, Innovation Centre, Carlow Institute of Technology, Carlow, Ireland
[4]Engineering Services, Office of Public Works, Dublin, Ely Place, Dublin 2, Ireland

Abstract

Integrated constructed wetlands (ICWs) can provide a management option for the collection and treatment of farmyard dirty water. An integrated constructed wetland is a surface flow wetland, which mimics the role and structure of natural wetlands. It also encompasses elements of landscape fit, biodiversity and habitat enhancement into its design. As part of the Annestown stream restoration project, ICWs were installed to treat dirty water from 12 farmyards. A detailed monitoring programme was conducted between 2001 and 2004. A significant reduction in organic matter, nutrients and faecal indicator bacteria between influent and effluent samples was recorded. There was no summer discharge from any of the farm wetland systems. An improvement in the quality of the receiving water (Annestown stream) was observed. A site-specific study suggested that P exported from an ICW system was similar to background levels of P export rates from land to water. In general, the ICWs monitored during this study indicate that they are capable of treating farmyard dirty water to a satisfactory quality, thereby provide a means of reducing nutrient and contaminant loss from agriculture.

Keywords: integrated constructed wetlands, farmyard dirty water, phosphorus, nitrogen surface water, groundwater.

Introduction

Water quality status of the Annestown stream, Co. Waterford, Ireland was classified by the Irish Environmental Protection Agency (EPA) as heavily polluted during the 1990s. Dirty water runoff from farmyards within the watershed was considered to be a significant contributing factor to this. To help improve the water quality of the Annestown stream, 12 Integrated Constructed Wetlands (ICWs) were designed and built to treat farmyard dirty water from farms within the 25 km^2 Anne valley watershed. An ICW is a surface flow wetland, which mimics the role and structure of natural wetlands. The wetland is designed to have landscape fit, and by its addition to the landscape, it provides an additional ecological habitat (Harrington and Ryder, 2002). The design and structure of these systems are described in more detail by Harrington *et al.* (2005). The initiative to adopt an ICW approach within the watershed began in 1996. Most ICWs were built and commenced operation between 1999 and 2000.

This paper presents an overview of a monitoring study of 12 farm ICWs during the last three years, and their impact on receiving waters (both surface and groundwaters) within the watershed.

Materials and methods

Study area

Wetlands were constructed on various farm enterprises within the watershed (Table 1). In general, wetland size ranged from 3,621 m^2 to 22,025 m^2. Components of farmyard dirty water discharged to the wetlands were variable; waters typically consisted of yard and dairy washings, rainfall on open yard and farmyard roofed areas along with silage and manure effluents. On average, total ICW size was about 1.4 times the size of the open farmyard areas. All ICWs are in operation for at least four years.

Water sampling and flow monitoring

An extensive water quality monitoring programme was undertaken on the 12 ICW systems. This comprised of sampling wetland influents, effluents, receiving surface waters and groundwaters between August 2001 and July 2004. Grab samples were taken on a monthly basis from each wetland influent and effluent during the monitoring period. Wetland influents were generally taken from the first cell of each ICW, while effluent samples were taken from either the outlet pipe of the last wetland cell, or in the absence of effluent discharge, from the surface waters of the last wetland cell of each ICW system.

At one farm, ICW number 11 (a 4-celled system)flow meters were installed. Flows into (influent), between wetland cells (inflow into cell 2) and out of the ICW (effluent) were continuously recorded, using a flow meter and recording device. Flows were measured between March 2003 and July 2004.

Monthly grab water samples were taken since August 2001 along the main channel of the Annestown stream. This stream is the receiving waterbody for all ICW discharges. Macroinvertebrate surveys of the stream were also taken every three years by the EPA, which is an independent public body that was set up to protect the environment.

A total of eleven piezometric groundwater-monitoring wells were placed both up gradient and down gradient at various depths (2-5 meters) at three of the farm ICW sites. These wells were sampled approximately monthly between January 2004 and July 2004. The day before sample extraction, wells were purged. To assess the status of ambient groundwater quality, water samples were also taken from the farmers' water-supply wells. Testing of the participating farmers' water-supply wells was conducted once during March 2004.

Laboratory analysis

Water analysis for chemical oxygen demand (COD), five day biological oxygen demand (BOD_5), ammonium, nitrate, molybdate reactive phosphate (MRP), total suspended solids (TSS), chloride

Table 1. Site characteristics of farms and integrated constructed wetland systems in the Anne valley, Co. Waterford.

ICW number	Farm enterprise	Farmyard area (m²)	Effective ICW area (m²)	ICW to yard area ratio	Aspect ratio of wetland cells	Dairy washings[1] (number of cows)	Yard water	Silage effluent	Roof water	Extraneous surface water or spring
1	Dairy	4,500	6,081	1.4	3	Yes (60)	Yes	No	Yes	Yes
2	Dairy	14,750	22,025	1.5	4	Yes (60)	Yes	Yes	Yes	Yes
3	Dairy, Beef	5,400	10,290	1.9	9	Yes (50)	Yes	Yes	Yes	No
4	Dairy	8,700	10,340	1.2	5	Yes (100)	Yes	No	Yes	No
5	Dairy and Tillage	4,000	3,950	1.0	2	Yes (35)	Yes	Yes	Yes	Yes
6	Dairy	9,800	12,710	1.3	2	Yes (80)	Yes	Yes	No	Yes
8	Beef	2,300	3,942	1.7	2	No	Yes	No	Yes	Yes
9	Mixed	4,800	7,958	1.7	2	Yes (55)	Yes	Yes	Yes	No
10	Mixed	2,100	4,380	2.1	2	Yes (50)	Yes	Yes	Yes	No
11	Dairy	5,000	7,690	1.5	5	Yes (77)	Yes	Yes	Yes	No
12	Mixed	10,500	10,726	1.0	2	Yes (85)	Yes	No	Yes	No
13	Sheep and tillage	5,000	3,621	0.7	2	No	Yes	No	Yes	Yes

[1] Dairy washings: A "Yes" indicates discharges to the wetland; a "no" indicates that there were none.

and *E. Coli* bacteria was conducted at the Waterford County Council water laboratory using American Public Health Association (APHA) standard methods 1992. The analysing laboratory participates in the EPA's laboratory proficiency scheme and is registered as an approved laboratory for these water analyses.

Results

Wetland influents and effluent water quality

Water quality data for all wetland influents and effluents for the monitoring period August 2001 until July 2004 are summarised in Table 2. Concentrations of the various water quality parameters measured reduced between wetland inlets and outlets. Concentration reductions were generally greater than 96% independent of water quality parameter. On average, COD was 2200 ± 8000 mg l^{-1}, BOD$_5$ was 1040 ± 4800 mg l^{-1}, TSS was 677 ± 22 mg l^{-1}, ammonia was 83 ± 0.7 mg l^{-1}, MRP was 24 ± 0.5 mg l^{-1} and *E. Coli* was 216000 ± 381000 cfu per 100 mls at wetland influents. This suggests that farmyard dirty water that was discharged to the 12 wetlands was variable in water quality. Average effluent water quality being discharged from the wetlands was 51 ± 34 mg COD l^{-1}, 19 ± 20 mg BOD$_5$ l^{-1}, 22 ± 26 mg TSS l^{-1}, 0.7 ± 2.0 mg NH$_4^+$-N l^{-1}, 0.5 ± 0.8 mg MRP l^{-1}, and *E. Coli* was 368 ± 1320 cfu per 100 mls during monitoring period.

In terms of BOD$_5$, if the two farm sites without dairy washings are excluded, ICWs 8 and 13 (Table 1) then the BOD$_5$ concentration of water entering the other ICW systems was approximately 1200 ± 5200 mg l^{-1}. This value is consistent with soiled water, such as dairy washings diluted with "cleaner" water such as yard and roof runoff (EPA, 2002).

The ICW systems seem particularly proficient at reducing nitrogen concentrations in farmyard dirty water. Although not all fractions of nitrogen, such as organic nitrogen, were measured in this study, it is reasonable to assume that the sum of ammonium and nitrate in ICW system effluent are the major components of nitrogen in the treated water, given the low particulate and organic matter content in wetland effluents (Table 2). All phosphorus analyses were on the unfiltered reactive fractions, which are represented as MRP. During the monitoring a sub-set of wetland effluent samples were analysed for total phosphorus (TP). Results indicate that the effluent MRP concentrations ranged between 60 and 80% of TP concentrations.

There was a marked reduction in *E. Coli* counts between influents and effluents (Table 2). In fact, *E. Coli* counts in the ICW effluents being discharged to surface waters were lower than background counts in other south-eastern river waters, which had an average of 1446 cfu per 100 mls for the years 2002 and 2003 (Neill, 2002; 2003). This suggests that the ICWs had a high capacity to remove faecal indicator organisms associated with farmyard dirty water.

Factors influencing wetland performance

While a full discussion of ICW design in relation to farmyard management practices is beyond the scope of this paper, we will discuss the relative sizing of ICWs to farmyard areas, and also wetland aspect ratios, in conjunction with phosphorus performance data of the wetland. The average MRP concentration in the ICW effluents is used as the key indicator of water treatment performance. Effluent MRP concentration is considered an appropriate indicator, as phosphorus is recognised as one of the most difficult nutrients to remove from water, and it is the limiting nutrient in most Irish freshwater systems. To do this determination of performance six ICW systems (3, 4, 9, 10, 11 and 12; Table 1) were considered. These sites were chosen as on-site observations of all wetlands suggested that the other sites (1, 2, 5, 6, 8, and 13) were being influenced by the entry of extraneous waters from surface and/or subsurface origins. Due to these confounding effects, these sites were excluded.

For the purpose of analysing the influence of ICW size on performance, the intercepted area from which water was observed to flow to the wetland was measured. Based on this, the ICW to intercepted area ratios ranged from 1.3 to 3.8. There was an inverse correlation (r^2=0.66; $p < 0.05$) between these ratios and wetland effluent MRP concentration (Table 2). These findings suggest that interception area such as farmyard open areas were important for adequate sizing of ICWs to attain wetland effluent MRP concentrations of sufficient quality.

Aspect ratio is defined as the average length of the wetland system divided by the average width. In this study, aspect ratios ranged from 2 to 9. There was a strong correlation (r^2=0.86; $p < 0.05$) between aspect ratio and wetland effluent MRP concentration.

There was no relationship between average wetland influent and effluent MRP concentrations. Nor was there a relationship between cow numbers and wetland effluent MRP concentrations. This may suggest that the amount of dairy wash water generated per cow has little effect on the overall performance of the ICW. Silage effluents contributed to farmyard dirty water on some sites. We found that there was no significant difference in effluent MRP concentration between systems receiving silage effluent, and those ICW systems that were not.

Table 2. Summary data of water quality parameter concentrations of ICWs influents and effluents. Water quality parameters include chemical oxygen demand (COD), five day biochemical oxygen demand (BOD_5), total suspended solids (TSS), ammonium (NH_4^+-N), molybdate reactive phosphorus ($PO_4^{3-}-P$), nitrate (NO_3-N), chloride (Cl^-) and E. Coli.

ICW number		COD (mg l⁻¹)		BOD_5 (mg l⁻¹)		TSS (mg l⁻¹)		NH_4^+-N (mg l⁻¹)		$PO_4^{3-}-P$ (mg l⁻¹)		NO_3-N (mg l⁻¹)		Cl^- (mg l⁻¹)		E. Coli (cfu per 100 mls)	
		Influent	Effluent	Influent	Effluent	Influent	Effluent	Influent	Effluent	Influent	Effluent	Influent	Effluent	Influent	Effluent	Influent	Effluent
1	Mean	10167	29	7194	12	1141	13	184	0.3	92.4	0.2		2.1	1687	54	39000	37
	SD	23261	11	16304	9	3113	12	258	0.4	220	0.4		2			61000	34
	n	21	24	19	23	21	22	21	22	20	23		23			3	6
2	Mean	789	38	510	31	532	24	64	0.5	15.3	0.3		5.7	187	51	122000	46
	SD	759	14	643	2	1639	33	63	0.8	7.7	0.2		3.7	1	1	187000	46
	n	19	24	14	23	18	22	18	20	18	24			1	1	4	3
3	Mean	1209	78	447	26	136	20	56	0.8	19.1	1.8		2	154	60	153000	87
	SD	1692	19	681	19	144	16	38	2.6	9.9	1.8		1.8	12	6	154000	20
	n	25	28	20	21	22	25	28	25	28	31		18	6	6	3	3
4	Mean	1897	96	521	34	1119	52	119	3.1	25.4	1.5		3.3	197	75	513000	373
	SD	1981	39	365	25	3180	40	63	4.6	11.2	1.1		3.5	32	7	600000	173
	n	33	32	27	26	31	29	34	31	35	34		24	7	7	3	3
5	Mean	922	43	313	19	136	14	51	0.6	13.6	0.3		3.8	178	27	190000	249
	SD	1505	34	382	22	131	9	68	1.1	22	0.7		2.1			41000	300
	n	21	20	21	20	21	18	19	20	20	21		23	7	7	3	3
6	Mean	594	55	213	19	210	22	39	0.3	10.5	0.1		1.8	1	35	23000	13
	SD	670	29	289	16	493	25	45	0.3	9.8	0.2		1.3			30000	6
	n	21	23	21	21	19	21	21	21	22	23		14			3	3
8	Mean	137	33	59	3	43	9	22	0.2	1.5	0.04		0.8	0	45	1000	79
	SD	87	9	62	14	60	7	85	0.3	1.3	0.07		0.6			1000	48
	n	22	22	20	19	21	20	21	21	22	22		10			3	3
9	Mean	828	44	494	12	459	15	41	0.6	11	0.5		1.2	259	41	94000	65
	SD	1522	14	1040	11	1025	11	46	1.1	22.2	0.3		1.1	136	4	110000	71
	n	25	24	25	24	25	21	31	31	31	32		4	7	7	3	3
10	Mean	834	43	171	14	165	20	40	0.4	14.7	0.1		1.7	48	23	24000	53
	SD	2171	35	398	11	267	10	108	0.8	36.8	0.1		1.5	19	3	20000	61
	n	26	27	22	23	23	23	28	31	27	31		22	5	7	3	2
11	Mean	1502	60	694	21	693	18	40	0.4	12.8	1		3.2	134	40	293000	10
	SD	2401	24	1066	21	2354	17	48	0.6	17.2	0.6		3.6	81	5	166000	
	n	40	41	32	28	37	39	43	46	43	47		21	10	12	3	2
12	Mean	6494	45	2245	17	3089	24	266	0.7	60.4	0.3		2.7	96	41	79000	1056
	SD	11995	56	2537	36	8340	28	407	1.8	80.2	0.5		2.3	101	4	95000	1374
	n	28	26	22	22	22	21	28	28	28	29		36	4	39	3	2
13	Mean	139	35	50	16	200	20	13	0.1	0.9	0.05		1.8	0	1	2000	816
	SD	186	18	70	15	370	22	26	0.1	1.3	0.08		1.3				1320
	n	18	19	17	17	18	17	18	19	17	19		9			1	3

All the ICWs have been operating for a minimum of four years, and their performance with time is illustrated in figure 1. Effluent MRP concentrations were variable between and within sites during the monitoring period. Molybdate reactive phosphorus concentrations ranged from 10 µg l^{-1} to 4.7 mg l^{-1} depending on site and time of year. ICW sites 3 and 4 had relatively high and variable effluent MRP concentrations, with mean effluent MRP concentrations of 1.8 and 1.5 mg l^{-1} respectively. The remaining sites displayed less variability and relatively lower (less than 1 mg l^{-1}) mean effluent MRP concentrations. Average MRP concentration for all wetland effluents was low (mean ± standard deviation; 0.41 ± 0.7 mg l^{-1}) in comparison to influent concentrations. There was no discernable disimprovement in effluent quality over the monitoring period.

Site specific hydraulic flows

Hydraulic influences from extraneous sources were minimal at site 11; thus flow meters were installed. Wetland inflows, inflows to wetland cell 2 and wetland outflows were continuously recorded and results are show in Figure 2. Average daily inflows, inflow to cell 2 and outflows from the ICW system during the monitoring period (April 2003 to August 2004) were 8.02 ± 4.54 m^3 d^{-1}, 10.25 ± 14.69 m^3 d^{-1}, and 0.62 ± 1.72 m^3 d^{-1}, respectively. Although there was inflow to the ICW between August 2003 and December 2003, and May 2004 and August 2004 there was no outflow from the system during these periods. The result being that there is a short discharge period to surface receiving waters. At this particular site discharge was during late winter and early spring. In general, all flows increased during early winter to early spring periods, which were from November to February 2004, whereas flows generally decreased during summer periods (Figure 2). During spring and early summer periods (February to May) inflows to wetland cell number two were larger than inflows, which suggest that rainfall on the previous wetland cell was an important contributor to within system flows. We can assume this, as the effect of other waters such as surface and subsurface runoff from the surrounding landscape was minimal at this particular site.

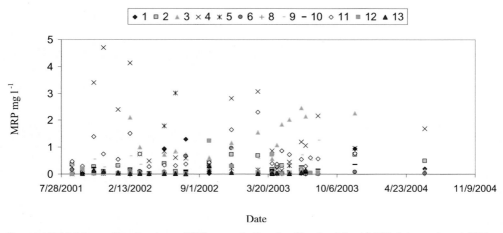

Figure 1. Molybdate reactive phosphorus (MRP) concentrations in effluents of the 12 ICWs between August 2001 and June 2004. Total number of observations during this period was 260.

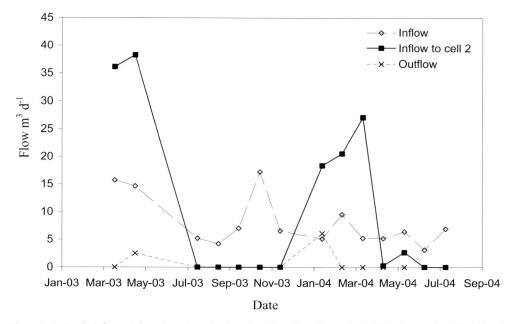

Figure 2. Hydraulic inflows, inflows to cell number 2 and outflows from the wetland (ICW, site number 11) during the monitoring period (April 2003 to August 2004).

The pattern of flows seen at site 11 was repeated at other ICW sites, where instantaneous flows were measured. During February 2004 an average increase in flow through the wetland ponds of 250 % was recorded at sites 4, 9 and 11. Decreases in flow through the wetlands were recorded in the summer periods, and none of the twelve farm ICWs were discharging to receiving waters during the period June to October 2004. During dry periods the loss of water from the wetland due to evaporation and infiltration creates increased water storage capacity or freeboard within wetland cells. This provides a buffering capacity of the wetland during wetter weather such that waters are stored within the wetland system prior discharge during wet weather.

Phosphorus export

We made comparisons between the export of P from the ICW site (site number 11) during the monitoring period and some other P export rates reported from other studies in Ireland that investigated P loss from agricultural practices to water. Exports of P from the ICW site were similar to background levels of P export from land to water and lower than P loads in runoff from agricultural grassland. Morgan *et al.* (2000) outlines that where farmyard dirty waters were not managed, leakage of P from farmyards can be high (Table 3). Thus, the appropriate management of these waters are important and our study indicates that the use of ICWs may be an appropriate management option.

Table 3. Comparison of phosphorus (P) exports from the ICW site number 11 and other land losses of P in Ireland.

Source	Phosphorus export rates kg P ha^{-1} yr^{-1}	Study
ICW site number 11, Anne valley, Waterford.	0.27 MRP	This study.
Farmyard leakage, Dripsey watershed study, Cork	7.95 as dissolved MRP	Morgan *et al.*, 2000
Background/natural land export	0.1 - 0.2 as TP	McGarrigle *et al.*, 1993
Agricultural grassland export	1 - 2 as TP	Tunney *et al.*, 1998

Receiving surface waters

Since the construction of the 12 farm-scale ICWs, the Annestown stream has had good chemical water quality status (Table 4). All four sites that we monitored along the stream complied with the target phosphorus concentration of a median annual concentration of 0.03 mg MRP l^{-1} as required by the Irish Phosphorus Regulations (1998). Biological water quality status was also good. Since 1998, the biological water quality status of the stream has improved from an overall water quality rating of Q2 (seriously polluted) to a water quality rating of Q 3/4 (slightly polluted) in 2001 and 2004.

Groundwater quality

All ICWs were constructed using *in situ* soils. Site subsoil was used and reworked to line wetland bed and bank surfaces and topsoil was redistributed for plant establishment. This helps to impede infiltration from wetland surfaces. In addition, incoming waters (farmyard dirty water)

Table 4. Annestown stream water quality monitored between 2001 and 2004. Values represent median, mean, standard deviation (SD) and number of observations (n) at each site.

	Sampling station (distance from sea)	mg L^{-1} BOD$_5$	Ammonia	MRP	Nitrate
Median (Mean)	Ballyphilip upstream of village (4 km)	1.3	0.05	0.02 (0.03)	4.3
SD		0.04	0.025	0.015	0.85
n		4	7	12	11
Median (Mean)	Ballyphilip downstream of village (3.5 km)	2	0.05	0.03 (0.03)	4.4
SD		1.6	0.03	0.02	0.81
n		18	13	22	28
Median (Mean)	Dunhill Castle (2 km)	1.9	0.04	0.02 (0.02)	5
SD		1.26	0.025	0.02	1.1
n		16	19	27	25
Median (Mean)	Monument (0.5 km)	2	0.06	0.03 (0.03)	4.2
SD		1.39	0.04	0.02	1.4
n		17	20	28	26

to the ICWs are high in BOD_5 and TSS loads. As these waters pass through the wetland, suspended organic material is typically deposited onto wetland soil surfaces, which helps impede infiltration from wetland cells (Kadlec and Knight, 1996).

Nitrate levels were elevated (> 10 mg l^{-1}) in most of the participating farmers' water-supply wells (Table 5), which were monitored to assess background nitrate concentrations in groundwaters. These high concentrations suggest that the farmers drinking water supply was impaired during the time of sampling. Nitrate concentrations of groundwaters in wells located upgradient, downgradient or beside ICW treatment wetland cells was generally lower (< 0.6 mg l^{-1}) than farm wells; however at ICW site one; concentrations were 11.7 mg l^{-1}, similar to ambient levels. Ammonium concentrations for all wells upgradient, downgradient or beside ICW treatment wetland cells other than those present at site three were 2.44 ± 2.12 mg NH_4^+ l^{-1}. At site three concentrations were much higher than all other wells. Ammonium concentrations of well waters at this ICW site averaged 18.67 ± 15.07 mg l^{-1}. A possible explanation for these high concentrations is that these wells are within the floodplain of the Annestown stream, which may have naturally high ammonium levels due to the waterlogged conditions. The MRP and *E. Coli* levels of well waters (farm wells and wells associated with ICW systems) were generally low (Table 5). One borehole at ICW site 3 had elevated *E coli* levels.

A major component of ICW design and operation is to eliminate the need for conventional land spreading of dirty water as several studies in Ireland (Richards *et al.*, 1999; Bartley, 2003; Rodgers *et al.*, 2003) have reported that land spreading of dirty water can lead to (either directly or indirectly) high nitrate levels in soil pore water and groundwater. Given this evidence and the evidence we have outlined in our monitoring, it seems that ICWs can provide an appropriate alternative management option for such waters.

Conclusions

Significant concentration reductions in organic material, suspended material, nutrients and faecal bacteria, between ICW influent and effluents of the 12 sites was observed during the monitoring study. Surface discharges from the ICW sites were seasonal. None of the 12 farm ICWs had surface discharges during summer months. The export of phosphorus from the intensively monitored site to surface waters was similar to background levels of P export rates.

Piezometer well water quality data suggest that ICWs had negligible effects on receiving groundwaters. This is important as present landspreading practices of farmyard dirty water can lead to the degradation of groundwater quality.

In conclusion, our findings suggest that ICWs are capable of treating farmyard dirty water to a satisfactory quality and that these systems can provide a suitable management option to reduce nutrient and contaminant loss from agriculture to water resources.

Table 5. ICW groundwater monitoring results between January and July of 2004.

	Site number	Location	Depth m	Ammonium mg l^{-1}	Nitrate mg l^{-1}	MRP mg l^{-1}	*E. Coli* per 100 mls
	1	Farm well		0.02	13.4	< 0.02	0
Mean		Well down gradient of ICW treatment cell 3	3	5.8	0.7	< 0.03	< 50
SD				5.4	2.4	0.03	
n				16	14	15	2
Mean		Well down gradient of ICW treatment cell 5	5	0.96	11.7	<0.02	< 50
SD				0.44	6.1	< 0.02	
n				5	5	5	1
	3	Farm well		0.01	10.4	< 0.02	0
		Farm well		0.004	11.9	< 0.02	0
Mean		Adjacent farmyard well	5	0.29	11	0.05	
SD				0.38	6.3	0.06	
n				6	6	6	
Mean		Well up gradient of ICW treatment cell 2	5	36	< 0.03	0.09	100
SD				25.6	0.13	0.09	
n				6	6	6	2
Mean		Well beside ICW treatment cell 3	5	11.3	1.6	< 0.03	>5000
SD				7.6	3	0.06	
n				14	12	14	2
Mean		Well down gradient of ICW treatment cell 4	3	8.7	1.3	0.17	75
SD				6.8	2.6	0.49	
n				5	5	5	0
	11	Farm well		0.01	12.6	< 0.02	6
		Farm well		0.01	9.9	< 0.02	1
Mean		Well up gradient of treatment cell 1	5	0.4	0.1	0.02	< 50
SD				0.17	0.11	0.006	
n				4	4	4	1
Mean		Well down gradient of treatment cell 3	5	0.77	< 0.02	0.01	< 50
SD				0.29	0.23	0.006	
n				7	7	7	2
Mean		Well down gradient of treatment cell 3	3	3.8	< 0.03	< 0.03	< 50
SD				1.6	0.18	0.018	
n				7	7	7	2
Mean		Well down gradient of treatment cell 4	5	2.9	< 0.03	< 0.02	< 50
SD				1.98	0.13	0.076	
n				7	7	7	2
Mean		Well down gradient of treatment cell 5	5	0.3	< 0.03	< 0.02	< 50
SD				0.2	0.18	0.008	
n				7	7	7	2

Acknowledgements

The authors wish to thank the participating farmers for their kind cooperation, Ann Gannon for providing rainfall data, and Sinead Fox and Declan Halpin, Waterford County Council, for their contribution to this study.

References

American Public Health Association, 1992. Standard Methods for the Examination of Water and Wastewater, 17th edition, APHA AWWA WEF, Washington, D.C.

Bartley, P., 2003. Nitrate responses in groundwaters under grassland dairy agriculture. Ph.D. thesis, Trinity College Dublin, Ireland.

EPA, 2002. Interim report: The biological survey of river quality 2001. Environmental Protection Agency, Johnston Castle, Wexford, Ireland.

Harrington, R., and C. Ryder. 2002. The use of integrated constructed wetlands in the management of farmyard runoff and waste water. In: Proceedings of the National Hydrology Seminar on Water Resource Management: Sustainable Supply and Demand. Tullamore, Offaly. 19th Nov. 2002. The Irish National Committees of the IHP and ICID, Ireland.

Kadlec, R.H. and R.L. Knight, 1996. Treatment Wetlands. CRC/Lewis Publishers, Boca Raton, FL. USA, 893 pp.

McGarrigle, M., T. Champ, R. Norton, P. Larkin, and M. Moore, 1993. The trophic status of Lough Conn, County Mayo, Ireland. Mayo County Council.

Morgan, G., Q. Xie, and M. Devins, 2000. Phosphorus export from farm in Dripsey catchment, Co. Cork, Ireland. EPA Research and Development Report, Series No. 6, pp. 17-25.

Neill, M., 2002. River water quality in the South East, 2002, Environmental Protection Agency, Johnston Castle, Wexford, Ireland.

Neill, M., 2003. River water quality in the South East, 2003, Environmental Protection Agency, Johnston Castle, Wexford, Ireland.

Richards, K. 1999. Sources of nitrate leaching to groundwater in grasslands of Fermoy, Co. Cork. Ph.D. thesis, Trinity College Dublin, Ireland.

Rodgers, M., P. Gibbons, and J. Mulqueen, 2003. Nitrate leaching on a sandy loam soil under different dairy wastewater applications. In: IWA Proc. 7th International Specialised Conference on Diffuse Pollution and Basin Management, UCD, Dublin, Ireland. Aug. 2003.

Tunney, H., O. Carton, T. O'Donnell, and A. Fanning, 1998. Phosphorus loss from soil to water. End of Project Report. Teagasc, Ireland, ARMIS 4022.

The relationship between plant vigour and ammonium concentrations in surface waters of constructed wetlands used to treat meat industry wastewaters in Ireland

A. Harrington
13 Ardfield View, Grange, Douglas, Cork, Ireland

Abstract

Constructed wetlands have the ability to treat various wastewaters with varying nutrient concentrations. Studies have been carried out to assess the impact of influent concentrations of ammonium on the growth of helophyte species in constructed wetlands used to treat wastewaters. Plants used in constructed wetlands have shown to have a threshold level of between 100 mg l^{-1} and 200 mg l^{-1} ammonium. The objectives of this study were to investigate the relationship between ammonium concentrations in wetland surface waters and plant vigour of a helophyte species *Carex riparia* (Curtis). This study was carried out on an integrated constructed wetland (ICW) treating waste water from a meat processing facility in Co. Carlow. The wetland comprised of three different segments. For the purposes of sampling each segment was divided into sections. At each sampling station the characteristics of plant growth were recorded and water samples were taken, which were later analysed for ammonium. Results suggest that as ammonium concentrations decreased with distance from wetland inlet down through the wetland system, plant vigour increased. Ammonium concentrations in upper wetland segments were 388 ± 131 mg l^{-1} (mean ± standard deviation), whereas in the lower section, average concentrations were 171 ± 41 mg l^{-1}. Plant growth seemed to be limited in upper wetland segments.

Keywords: Constructed wetlands, helophyte species, ammonium concentration, *Carex riparia* (Curtis).

Introduction

Constructed wetlands are designed to treat various types of wastewaters through a combination of physical, chemical and biological processes. Free surface flow constructed wetland treatment systems are designed to allow water flow above the wetland substrate. They are typically densely planted with helophyte species, which are plant species that are specifically adapted to survive and grow under a wide range of environmental conditions that can range from flooded to dry conditions. The ability of helophyte species to assimilate nutrients, to create favourable conditions for bio-films that enhance microbial decomposition of organic matter and increase residence time for through-flowing water are the main reasons they are often used in constructed wetlands to help improve water quality.

Plants can extract and use various forms of nitrogen (N) from soil porewater, most importantly are the inorganic ions ammonium (NH_4^+) and nitrate (NO_3^-). Although plants require ammonium for growth, studies have shown that there is a threshold level of ammonium, which plants can tolerate (McCaskey *et al.*, 1994). McCaskey *et al.* (1994) indicated that high ammonium concentrations (in excess of 100 mg l^{-1}) can be toxic to plants such as wetland vegetation. A study carried out by Clarke and Baldwin (2002) to assess the impacts of ammonium concentration on helophytic species in constructed wetlands that were treating animal waste, found that ammonium levels in excess of 200 mg l^{-1} inhibited the growth of *Juncus effusus* (L), *Sagittaria latifolia* (Willd) and *Typha latifolia* (L). The growth of *Schoenplectus tabernaemontani* was limited by ammonium concentrations in excess of 100 mg l^{-1}. Ammonium can be a substantial component in polluted wastewaters, such as meat processing industry wastewaters. Therefore, ammonium concentrations are often a fundamental limiting factor in the design of constructed wetlands to treat these waters.

Vegetation such as *Typha* and *Juncus* sp., are often the most obvious biological component of a constructed wetland ecosystem. The design, construction and maintenance of constructed wetlands is often focused towards the establishment of dense vegetation stands in order to optimise nutrient assimilation and provide suitable conditions for microbial decomposition processes (Peterson, 1998). However, in heavily nutrient loaded constructed wetland systems such those previously mentioned, the concentration of ammonium may limit vegetation establishment and consequently reduce nutrient assimilation by the wetland.

Under aerobic wetland conditions ammonium is transformed to nitrite and then to nitrate (nitrification), which is then denitrified to N_2 under anaerobic conditions. Helophytic vegetation has the ability to transfer oxygen from plant shoots to roots, which facilitates an oxidised micro-environment around plant roots and rootlets, which is known as a plant's "rhizosphere." This oxidised micro-environment stimulates decomposition processes and the growth of nitrifying bacteria.

The overall objective of this study was to determine the relationship between the concentration of ammonium in wetland surface waters and plant vigour, as defined by growth and re-growth of *Carex riparia* Curtis (greater pond sedge), which is a commonly used plant within constructed wetland systems used to treat meat industry wastewaters in Ireland.

Materials and methods

Site description

The site for the study was an 8000 m^2 integrated constructed wetland (ICW) treatment system built in 2002 to treat meat industry wastewater from Ballon Meats Ltd, Co. Carlow (Figure 1). It comprised of four ponds/segments all of which cover a similar area of about 2000 m^2. Wetland segments one to three were used in this study. A number of sample sites were used within each segment of the wetland to measure plant vigour and take samples of overlying water, which would be later analysed for ammonium concentrations.

The effluent from the meat factory was initially treated through dissolved air floatation (DAF) unit. The effluent from the DAF unit was then allowed to discharge to the first wetland. The wetland was designed to treat 150 m^3 of effluent a day, through the four wetlands. The influent ammonia concentrations were often in excess of 400 mg l^{-1}. In situ soils were used for wetland construction.

Wetland segments one to three were planted with *Carex riparia (Curtis), Typha latifolia* (L) and *Glyceria maxima (Hartm.)*. The plants were planted approximately one meter apart in order to achieve a dense vegetation cover throughout most of the wetland. *Carex riparia* was the dominant plant species used, comprising about 80% of the individual plants planted.

Species description

Carex riparia Curtis (Common Pond Sedge) is a perennial evergreen that grows to 1.5 m to 2 m. This species is usually glabrous, with three angled stems and rather harsh grass-like leaves (Figure 2). In Ireland, it flowers from May to June and seeds typically ripen between July and August. Naturally, it is found in ditches, marshes and river-banks, mainly near coastal areas, occasionally in the east of Ireland, but rare elsewhere. Plants can be easily cultivated in damp to wet soil in full sun or shade.

Figure 1. *Wetland segment one of constructed wetland system used to treat meat industry wastewaters at Ballon Meats, Co. Carlow.*

Figure 2. *Habit of Carex riparia in flower May to June.*

Sampling

A systematic sampling system was carried out to investigate the relationship between ammonium concentration and plant vigour. Sampling was carried out during June 2002. There were sixteen sampling stations within the first three wetland segments. The first two segments were divided into three sections; upper, middle, and lower. The third wetland segment was divided into upper and lower sections only.

At each sample station the overlying water was sampled and collected samples were analysed for ammonium. Parameters such as concentration of ammonia in the overlying water; height of the individual plant; diameter of the plant; number of new shoots on the plant; colour of plant; and the density of plants in the row were recorded.

Results

Results are shown in Table 1. Concentrations of ammonium were highest in wetland segment one in comparison to the other two wetland segments. Surface water concentrations of ammonium also decreased between upper, middle and lower sections of each wetland segment.

Plant height, plant diameter and the number of new shoots on plants were lower in wetland segment one, in comparison to the second and third wetland segments. These results may suggest that plants were stressed in wetland segment one and with distance from the inlet, plant stress, as assessed by measuring plant height, plant diameter and number of new shoots, decreased with increasing distance from influent (Figure 3). There was a significant relationship between the concentration of ammonium in wetland surface waters and the number of new shoots on wetland plants ($p < 0.05$).

Table 1. Surface water ammonium concentrations and plant vigour parameters of the integrated constructed wetland that was used to treat meat industry wastewaters. Values represent means ± one standard deviation.

Wetland segment	Sections	Ammonia (mg l^{-1})		Plant height cm		Plant diameter (cm)		New shoots on plant		Plant colour
1	Upper	630	156	47.5	3.5	1.75	0.4	1	1.4	light/grey
1	Middle	480	71	47.5	3.5	14	5.6	6	2.8	yellow/green
1	Lower	380	57	47.5	25	7.5	6.4	0	0	brown/green
2	Upper	305	21	76.5	9	18	1.4	1	1.4	dark/green
2	Middle	275	7	66.5	9	17	1.4	1.5	2.1	yellow/green
2	Lower	260	-	77.5	3.5	18	8.5	5.5	3.5	green/brown
3	Upper	212	18	51	47	12.5	4.9	2	0	green
3	Lower	130	14	110	7	30	5.6	8.5	0.7	dark green
Inlet		440								
Outlet		82								

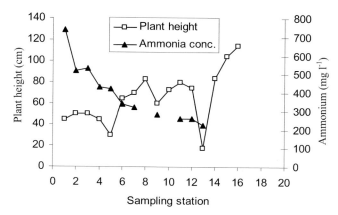

Figure 3. Plant height and ammonium concentration at each sampling station in the different wetland segments down through the constructed wetland system.

In general, the threshold concentration for ammonium in this constructed wetland system was about 200 mg l[-1]. Thus, these findings are somewhat similar to others reported even though there different wetlands and different helophyte species were investigated.

Conclusions

The study carried out indicated that while *Carex riparia* was found growing in effluent with a concentration in excess of 400 mg l[-1], its growth was severely inhibited. As ammonium concentrations decreased below 200 mg l[-1] plant vigour significantly increased, as there was an improvement in the physiological structure of the plant, as indicated by increased plant height, plant diameter and the number of new shoots.

When designing constructed wetlands for the treatment of concentrated effluents, such as meat industry wastewaters, it is important to reduce the concentration of influent ammonium to the wetland, as these concentrations seem to limit plant growth. An effective and simple way of reducing the concentration of ammonium discharging to a wetland would be to dilute the incoming effluent with on site storm water, generated from rainfall on site impervious surfaces.

References

Clarke, E. and A.H. Baldwin, 2002. Responses of wetland plants to ammonia and water level. Ecological Engineering. **18** 257-264.

McCaskey, T.A., S.N. Britt, T.C. Hannah, J.T. Eason, V.W.E. Payne and D.A. Hammer, 1994. Treatment of swine lagoon effluent by constructed wetlands operated at three loading rates. pp. 23-33. In: Constructed wetlands for animal waste management, edited by P.J Du Bowy and R.P Reaves, Purdue Research Foundation, West Lafayette, IN.

Peterson, H.G., 1998. Use of constructed wetlands to process agricultural wastewater. Canadian Journal of Plant Science **78** 199-210.

A small-scale constructed wetland to treat different types of wastewaters

G. Garcia-Cabellos[1], M. Byrne[1], M. Stenberg[1], K. Germaine[1], D. Brazil[1], J. Keohane[2], D. Ryan[1] and D.N. Dowling[1]

[1]Department of Science & Health, Institute of Technology Carlow, Kilkenny Rd., Carlow, Ireland
[2]GES ltd, Innovation Centre, IT Carlow campus, Kilkenny Rd., Carlow, Ireland

Abstract

Constructed wetlands are man-made systems that can be designed for the treatment of polluted water. One of the main mechanisms of detoxification is considered to be the high microbial activity in the nutrient and oxygen rich environment surrounding wetland plant roots. A simple small-scale wetland model system was developed to evaluate the ability of a constructed wetland system to treat three types of wastewaters. These were skimmed milk, raw sewage or water containing high concentrations of zinc. Our results showed a reduction of all the physico-chemical parameters examined with the exception of phosphates. Extracellular protease producing bacteria were found at higher counts in wetland soils and plants. We conclude that our small-scale system can be used to evaluate some of the microbial parameters of constructed wetlands.

Keywords: microcosm, organic waste, sewage, zinc, constructed wetlands.

Introduction

Constructed wetlands are efficient and cost effective methods for the treatment of waste water contaminated with high levels of nitrates, phosphates, sulphates, biological oxygen demanding (BOD), and suspended solids (Reed and Brown, 1995). They have been used to treat municipal wastewater (Kemp and George, 1997; White *et al.*, 2000), treat acid mine drainage (Mitsch and Wise, 1998), remove crude oil (Groudeva *et al.*, 2001), and to treat a wide range of metals (Wood and McAtamney, 1996, Webb *et al.*, 1998) and xenobiotics present in waters (Cheng *et al.*, 2002).

The aim of this work was to construct a simple small scale-scale laboratory model wetland system and evaluate its ability to treat organic wastes and heavy metals. This could be used to test plants and microbial inoculants in advance of wetland construction. Mungur *et al.* (1997) described a similar approach using larger size tanks for the removal of heavy metals.

Materials and methods

Constructed wetland microcosms

A model wetland system (microcosm) was constructed in the laboratory as shown in Figure 1a. The system consisted of a glass tank (1 m x 0.35 m x 0.30 m) and soil subsrate, which was a mix of perlite and vermiculates, to form the bed of the microcosm. Gravel was placed at the bottom corners of the microcosm and the soil mix was placed between them. The input end of the tank was raised slightly to produce a slope of 1% to aid the flow of liquid though the wetland. This channelled the waste water from the tank into collection containers for analysis (Figure 1a). The system received natural sunlight and the residence time was calculated as 27.6 days. All the plants used for this experiment were sourced from a constructed wetland in Kilmeaden, Co Waterford. They were planted as shown in Figure 1b.

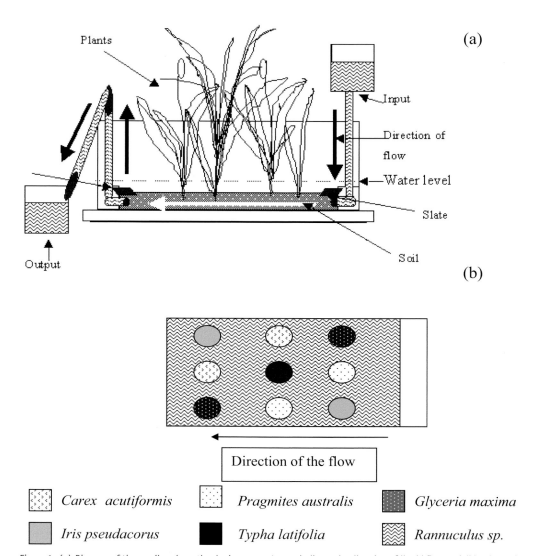

Figure 1. (a) Diagram of the small-scale wetland microcosm. Arrows indicate the direction of liquid flow and (b) schematic diagram showing the planting location within the microcosm.

Addition of wastewaters

Three different types of wastewaters, in three separate experiments, were added to the wetland:
- One litre of skimmed milk was added every two days to each system through the input stream. This was eventually reduced to one litre per week at which point sampling was carried out.
- One litre of raw sewage containing 2×10^5 CFU ml^{-1} of coliforms was added to the system through the input chamber and let flow through. The total number of coliforms was determined by the MPN method for a period up to eight weeks.
- Water spiked with zinc at a concentration of 240 mg l^{-1} was added to the microcosm. Soil and water samples were taken to determine the ability of the microcosm to retain Zn. Zinc concentrations were determined by atomic absorption spectroscopy.

The microcosm was sampled weekly at three sample points: input, output, and main tank. Soil, water and roots were also sampled from the main tank. Physico-chemical and microbial parameters were analysed using standard methods (Clesceri *et al.*, 1989).

Results and discussion

Skimmed milk

Table 1 shows the concentration reductions between wetland input and output to and from the wetland microcosm after the addition of skimmed milk. The system performed well during the test period of 10 weeks. The majority of the parameters tested were found to be within the levels permissible as set out by the Urban Waste Water Treatment (UWWT) regulation (S.I. 254 of 2001) (Smith *et al.*, 2003) with the exception of BOD_5 which exceeded the maximum admissible concentration. The slight increase in phosphates probably indicated that the system was saturated due to the high nutrient load (White *et al.*, 2000).

The overall reduction in nitrates, sulphates, TDS, BOD, conductivity and turbidity show that the microcosm was successful in treating the organic wastewater. A secondary treatment would

Table 1. Physical-chemical analyses of the input/output of the wetland microcosms.

Parameter	Raw wastewater	Standard deviation	Treated wastewater	Standard deviation	Total reduction %
Nitrate (mg l^{-1})	15.97	3.22	6.84	3.17	57%
Phosphate (mg l^{-1})	2.90	0.48	3.72	0.8	28% increase
Sulphate (mg l^{-1})	22.0	1.34	1.3	0.42	94%
TDS (mg l^{-1})	637	35.61	39	11.62	94%
BOD (mg l^{-1})	2396	110.24	376	69.86	84%
Conductivity µS	933	56.05	186	27.35	80%
Colour Pt-Co	5411	1482.83	97	64.07	98%
pH	6.94	0.03	7.61	0.35	0.67

be recommended to fully treat the wastewater to achieve all water quality criteria are required for effluents.

Raw sewage

Total viable counts of extracellular protease degraders were carried out every week using skimmed milk media (SMA) and the results are detailed in the Figure 2.

Figure 2 shows that during the experimental period, the number of bacteria were always higher in soils and roots. This area, known as the rhizosphere, consists of a narrow soil zone surrounding the root. Within these areas interactions between plant and bacteria can occur. The root provides oxygen and organic matter for the bacterial population (O'Gara et al., 1994). The lowest numbers of bacteria were present in the water, possibly due to the lack of nutrients in the waters and the constant flow of liquid within wetland microcosm.

Initial coliform levels in the input were in the order of 10^5 CFU ml^{-1}, while levels in the output never rose above 10^2. This showed a 99.5% reduction rate of coliforms from the raw sewage. Karim et al. (2004) explained that one of the mechanisms for reduction of pathogens in constructed wetlands is sedimentation. The authors concluded that sedimentation of pathogens in vegetated wetlands appeared to be related to the higher sedimentation of larger particles.

Water spiked with zinc

Zinc concentrations were measured in the input, central and output areas of the microcosm (Figure 3). The ability of constructed wetlands to remove Zn has been described previously (Crites et al., 1997, Mays and Edwards, 2001, Cheng et al., 2002 Walker and Hurl, 2002). This experiment indicated that in a small laboratory microcosm with a diversity of plants, Zn concentrations can be efficiently reduced.

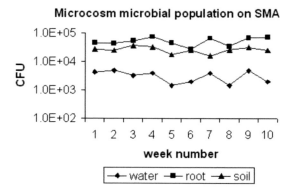

Figure 2. Change in total viable counts (CFU) of extracellular protease producers with time.

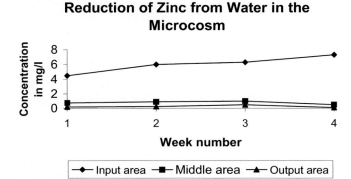

Figure 3. Zinc removal in the microcosm during the four weeks of monitoring.

Conclusion

- The laboratory small-scale constructed wetland system showed reductions for all of the physico-chemical parameters tested, except for phosphates.
- The total count of bacteria on skimmed milk agar (SMA) shows that the extracellular protease producing bacteria became well established in wetland soil and roots.
- Zinc concentrations decreased from input to output in the system, probably due to sedimentation (Mays and Edwards, 2002).
- The total number of coliforms decreased from wetland input to wetland output.

Acknowledgements

We would like to thank to Rory Harrington for supplying all the plants used in this experiment. The work was funded in part by Enterprise Ireland ARP programme, HEA PRTLI, TSR Strand 3 and European Union (EU) contracts QLK-3-CT2000-00164 and QLRT-2001-00101.

References

Cheng, S., W. Grosse, F. Karrenbrock, and M. Thoennessen, 2002. Efficiency of constructed wetlands in decontamination of water pollutes by heavy metals. Ecological Engineering **18** 317-325.

Cheng, S., Z. Vidakosvic-Cifrek, W. Grosse, and F. Karrenbrock, 2002. Xenobiotic removal from polluted water by a multifunctional constructed wetland. Chemosphere **48** 415-418.

Clesceri, L., A. Greenberg, and R. Trussell, 1989. Standard methods for the examination of water and wastewater, 17th edition. American Public Health Association, American Water works Association and Water Pollution Control Federation.

Crites, R.W., G.D. Dombeck, R.C. Watson, and C.R. William, 1997. Removal of metals and ammonia in constructed wetlands. Water Environment Research **69** 132-135.

Groudeva, V.I., S.N. Groudev, and A.S. Doycheva, 2001. Bioremediation of waters contaminated with crude oil and toxic heavy metals. International Journal of Mineral Processing **62** 293-299.

Karim, M.R., F.D. Manshadi, M.M. Karpiscak, and C.P. Gerba, 2004. The persistence and removal of enteric pathogens in constructed wetlands. Water Research **38** 1831-837.

Kemp, M.C., and D.B. George, 1997. Subsurface flow constructed wetlands treating municipal wastewater for nitrogen transformation and removal. Water Environment Research **69** 1254-1262.

Mays, P.A and G.S. Edwards, 2001. Comparison of heavy metals accumulation in a natural wetland and constructed wetlands receiving acid mine drainage. Ecological Engineering **16** 487-500.

Mitsch, W.J., and K.M. Wise, 1998. Water quality, fate of metals, and predictive model validation of a constructed wetland treating acid mine drainage. Water Research **32** 1888-1900.

Mungur, A.S., R.B.E. Shutes, D.M. Revitt, and M.A. House, 1997. An assessment of metal removal by a laboratory scale wetland. Water Science Technology **35** 125-133.

O'Gara, F., D.N. Dowling, and B. Boesten, 1994. Molecular ecology of rhizosphere microorganisms. VCH Verlagsgesllschaft mbH Weinheim (Federal Republic of Germany) ISB3-527-30052-X.

Reed, S.C., and D. Brown, 1995. Subsurface flow wetlands- a performance evaluation. Water Environment Research **67** 244-248.

Smith, D., N. O'Neil, F. Clinton, and M. Crowe, 2003. Urban Waste Discharge in Ireland. A report for the years 2000 and 2001. Environmental Protection Agency (EPA), Wexford, Ireland.

Walker, D.J. and S. Hurl, 2002. The reduction of heavy metals in stormwater wetland. Ecological Engineering **18** 407-414.

Webb, J.S, S. McGuinness, and H.M. Lappin-Scott, 1998. Metal removal by sulphate-reducing bacteria from a natural and constructed wetlands. Journal of Applied Microbiology **84** 240-248.

White, J.S., S.E. Bayley, P.J. Curtis, 2000. Sediment storage of phosphorus in a northern prairie wetland receiving municipal and agro-industrial wastewater. Ecological Engineering **14** 127-138.

Constructed wetlands for wastewater treatment in Europe

J. Vymazal
ENKI o.p.s., Dukelská 145, 379 01 Trebon, Czech Republic

Abstract

The first experiments using wetland vegetation (macrophytes) for wastewater treatment were carried by out by Käthe Seidel in Germany in the early 1950s. The first full-scale free water surface constructed wetlands (FWS CWs) for wastewater treatment were put into operation in the Netherlands during 1967 and Hungary during 1968. Free water surface CWs did not spread widely in Europe and CWs with sub-surface flow have become more popular. The horizontal sub-surface flow systems were initiated by Seidel and improved by Reinhold Kickuth under the name Root Zone Method in late 1969 and early 1970s. The use of these systems spread throughout Europe in the 1980s and 1990s. These systems use porous substrates with high hydraulic conductivity for wastewater filtration. Vertical flow constructed wetlands (VF CWs) were also originally proposed by Seidel in the late 1960s as part of pre-treatment process for horizontal flow systems, which were used at full-scale in the mid-1970s and were called infiltration fields. Vertical flow systems were not used as much as horizontal flow systems because of higher maintenance and operation costs, but were used to nitrify wastewaters. Constructed wetlands have been used for many types of wastewater including industrial, agricultural, landfill leachate and stromwater runoff. As many of these wastewaters are difficult to treat in a single stage wetland system, hybrid wetland systems which consist of various types of constructed wetlands staged in series have been used in Europe.

Keywords: free water surface, horizontal flow, macrophytes, vertical flow, sewage.

Introduction

Constructed wetlands (CWs) are engineered systems that have been designed and constructed to utilise the natural processes involving wetland vegetation, soils, and the associated microbial assemblages to assist in treating wastewaters. They are designed to take advantage of many of the same processes that occur in natural wetlands, but do so within a more controlled environment. Constructed wetlands for wastewater treatment may be classified according to the life form of the dominating wetland vegetation (macrophyte) into systems with free-floating, rooted emergent and submerged macrophytes (Brix and Schierup, 1989). Most systems constructed in Europe are planted with emergent macrophytes, but the design of systems in terms of media, as well as the flow regime varies. The most common systems are designed with horizontal sub-surface flow (HF CWs), but vertical flow (VF CWs) systems are becoming more popular. Constructed wetlands with FWS are not used as much as the HF or VF systems despite being one of the oldest systems in Europe (Brix, 1994; Vymazal *et al.*, 1998; Vymazal, 2001). Constructed wetlands have been used for a long time mostly for the treatment of domestic or municipal sewage. However, recently CWs have been used for many other types of wastewater including industrial and agricultural wastewaters, landfill leachate or stromwater runoff. As many of these wastewaters are difficult to treat in a single stage system, hybrid systems, which consist of various types of constructed wetlands staged in series, were introduced in Europe.

First experiments with the use of macrophytes for wastewater treatment

The first experiments aimed at the possibility of wastewater treatment using wetland plants was undertaken by Käthe Seidel in Germany in the early 1950s at the Max Planck Institute in Plön (Seidel, 1955). Between 1952 and 1956, Seidel carried out numerous experiments on the use of wetland plants for treatment of various types of wastewater, including phenol containing wastewaters (Seidel, 1955, 1965a, 1966), dairy wastewaters (Seidel, 1976) and livestock wastewater (Seidel, 1961). In 1965, she carried out the first experiments with the use of plants for sludge dewatering (Bittman and Seidel, 1967). In early 1960s, the industrial cities along the Rhine River had to meet their water demand by soil filtration of the river water. To further clean the polluted water, the water-works tried to seep the water underground so that it could be mixed with clean groundwater. However, due to the different physical properties of surface water and groundwater they did not mix properly. Seidel recommended to pump water from the river and treat it in an aeration pond and then let it pass through the rhizosphere of bulrushes (*Schoenoplectus lacustris*), then to the subsoil, where it mixed with ground water (Czervenka and Seidel, 1965). This actually was a basis for vertical flow constructed wetlands.

In early 1960s, Seidel intensified her trials to grow macrophytes in wastewater and sludge of different origins and tried to improve the performance of rural and decentralised wastewater treatment, which was either septic tanks or ponded systems that were ineffective. She planted macrophytes into the shallow embankment of tray-like ditches and created artificial trays and ditches grown with macrophytes. Seidel named this early system the hydrobotanical method. Then she improved her hydrobotanical system by using sandy soils with high hydraulic conductivity substrates in a sealed module type basins planted with various macrophytic species. To overcome the anaerobic septic tank systems, she integrated a stage of primary sludge filtration in vertically percolated sandy soils planted with *Phragmites australis*. Thus, the system consisted of an infiltration bed through which, sewage flowed vertically and the discharge bed was a horizontal flow system (Seidel, 1965b). This system was a basis for hybrid systems that were revived at the end of the 20th century. Seidel´s concept to apply macrophytes to sewage treatment was difficult to understand for sewage engineers who had eradicated any visible plants on a treatment site for more than 50 years (Börner *et al.*, 1986) and therefore, it was no surprise that the first full-scale constructed wetlands were built outside Germany.

Free water surface constructed wetlands

In spite of many prejudices among civil engineers about odour nuisance, attraction of flies, poor performance in cold periods the IJssel Lake Polder Authority in Flevoland, Netherlands constructed its first FWS wetland in 1967 (de Jong, 1976; Veenstra, 1998). The wetland was planted with *Schoenoplectus lacustris* and had a design depth of 0.4 m with a total area of one hectare. A star shape layout was chosen in order to obtain optimum utilization of the available area; however this complicated macrophyte harvesting and maintenance in general (de Jong, 1976). To facilitate mechanical biomass harvesting and systems management longitudinal channels were added (Figure 1). The new wetland design included channels of 3 m wide and 200 m long, separated by parallel stretches of 3 m resulting in an increase in land requirement from 5 m^2 per one population equivalent (PE) for the star arrangement to 10 m^2 PE^{-1}. The system exhibited a very good treatment effect and in early 1970s, about 20 FWS of this ditch type, called planted sewage farm (or Lelystad process) were in operation in the Netherlands (Greiner and de Jong, 1984; Veenstra, 1998).

In 1968, FWS CWs were created in Hungary near Keszthely in order to preserve the water quality of Lake Balaton and to treat wastewater of the town (Lakatos, 1998). The constructed wetland was established in place of an existing natural wetland in peat soil. The existing natural wetland consisted of six ponds 40-60 cm deep with a surface area of 10 ha and the ponds were loaded with 8000 m^3 d^{-1} of mechanically pre-treated wastewater. In Hungary, FWS CWs were also created in 1970s to treat wastewaters from petrochemical industry. The FWS system in Nyirbogdány Petrochemical Plant has an area of 15,500 m^2 with *Typha angustifolia*, *Typha latifolia*, *Phragmites australis* and *Schoenoplectus lacustris* being the major emergent species and *Potamogeton pectinatus* being the major submerged species together with macroalga *Chara* sp. (Lakatos *et al.*, 1996; Lakatos, 1998). Other systems for petrochemical wastewaters used submerged vegetation (e.g., *Ceratophyllum demersum*) or filamentous algae (e.g. *Cladophora*

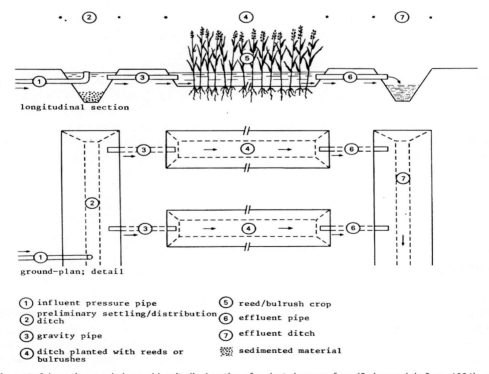

longitudinal section

ground-plan; detail

① influent pressure pipe
② preliminary settling/distribution ditch
③ gravity pipe
④ ditch planted with reeds or bulrushes

⑤ reed/bulrush crop
⑥ effluent pipe
⑦ effluent ditch
▨ sedimented material

Figure. 1. Schematic ground-plan and longitudinal section of a planted sewage farm (Greiner and de Jong, 1984).

glomerata). Contrary to North American approaches, FWS CWs did not spread throughout Europe significantly, while CWs with sub-surface flow drew much more attention at the end of the 20th century (Vymazal *et al.*, 1998). However, FWS CWs are in operation in many European countries, e.g. Sweden, Poland, Estonia and Belgium. In Sweden, FWS systems were constructed with nitrogen removal as a primary goal. Sometimes, the aim of FWS systems was to provide phosphorus polishing after chemical treatment and a buffer in case of conventional treatment plant failure (Sundblad, 1998). In most cases, FWS CWs in Sweden are used as tertiary treatment after conventional treatment. Similarly in Poland FWS systems are used for tertiary treatment and are quite large (Kowalik and Obarska-Pempkowiak, 1998). In the United Kingdom, FWS CWs are successfully used to treat highway runoff (Cooper *et al.*, 1996). At present, smaller FWS CWs are often used as a part of hybrid systems. Free water surface CWs are efficient in removing organics and suspended solids but nutrient removal is lower, usually about 50% (Table 1). However, removal efficiency expressed as a percentage is affected by the fact that many FWS systems are used for tertiary treatment and it is well known that the lower inflow concentrations the lower percentage removal.

Some emergent macrophytes are capable of forming floating mats, even though their individual plants are not capable of such existence. Some treatment systems have been operated in this fashion around the world and in some cases the floating mats developed unintentionally (Vymazal, 2001). However, constructed wetlands with floating mats of emergent macrophytes have recently been used successfully in Europe. These systems are, for example, used for treatment of runoff and de-icing waters at London-Heathrow airport or at Bornem in Belgium for treatment of street runoff.

Table 1. Treatment efficiency of FWS wetlands (Data from 85 systems in Australia, Canada, China, Netherlands, New Zealand, Poland, Sweden and USA).

	Average concentration (mg l^{-1})			Average loading			
	In	Out	% Reduction	In	Out	Removed	%
				(kg ha^{-1} d^{-1})			
BOD$_5$	34	9.5	70.3	12	4.3	7.7	68.4
TSS	53	14.4	72.9	16.2	4.7	11.5	71.0
				(g m^{-2} yr^{-1})			
TN	14.3	8.4	51.8	466	219	247	52.9
TP	4.2	2.15	48.8	268	136	132	49.1

Constructed wetlands with horizontal sub-surface flow

The most widely used design of CWs in Europe is horizontal sub-surface flow (HF CWs, Figure 2). The design typically consisted of a rectangular bed planted with the common reed (*Phragmites australis*) and lined with an impermeable membrane. Mechanically pre-treated wastewater is fed in at the inlet and flows slowly through the soil medium under the surface of the bed in a more or less horizontal path, until it reaches the outlet zone, where it is collected before discharge via water level control at the outlet. During the movement of wastewater through

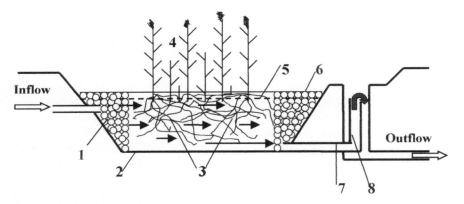

Figure 2. Schematic representation of a constructed wetland with horizontal sub-surface flow. 1-distribution zone filled with large stones, 2-impermeable liner, 3-filtration medium (gravel, crushed rock), 4-vegetation, 5-water level in the bed, 6-collection zone filled with large stones, 7-collection drainage pipe, 8-outlet structure for maintaining of water level in the bed. The arrows indicate only a general flow pattern (modified from Vymazal, 2001).

the reed bed, the wastewater comes into contact with a network of aerobic, anoxic and anaerobic zones.

This concept was developed in the late1960s in Germany by Käthe Seidel in Plön (Seidel, 1965). Seidel designed the system with filtration material with high hydraulic conductivity. However, Reinhold Kickuth from Göttingen University developed another system under the name Root-Zone Method (Kickuth, 1969, 1977). Kickuth´s system differed from Seidel´s system in the use of cohesive soils with high clay content. The first full-scale HF Kickuth´s type CW for treatment of municipal sewage was put in operation in 1974 in the community Liebenburg-Othfresen (Kickuth, 1977, 1978, 1981). An area of about 22 ha was originally used to dump waste material (silt, clay and dross) derived from mining of iron ore. It contained settlement ponds for separation of silk and clay, which were filled with clay, gravel, and chalk when mining ceased in 1962. In 1969, the local authority proposed to use part of the area for sewage treatment. The anaerobic maturation ponds were rejected and R. Kickuth recommended the RZM to be constructed (Boon, 1986).

Kickuth´s concept was closer to the traditional understanding of soil treatment for sewage, but his statement that root and rhizome growth would improve hydraulic conductivity of heavy soils failed on several early construction sites and it gave a harsh setback to scientific and official acknowledgement of HF CWs as a legitimate sewage treatment system. The problem in the constructed beds where soil was used was that as a medium they had low hydraulic conductivity, thereby resulting in surface flow, which resulted in a short retention time within the system. Kickuth (1977) also proposed the size of vegetated beds of only 2 m^2 PE^{-1}. It proved to be too small to achieve a satisfactory treatment effect. In 1983, German ideas were introduced in Denmark (Brix, 1987). Despite problems with surface flow, soil-based systems exhibited high treatment effect for most parameters if a reed bed area of 3-5 m^2 PE^{-1} was used. In order to overcome the overland flow, Danish systems were designed with a low aspect ratio (length:width ratio). It resulted in very wide beds and short passage length (Brix, 1998). However, the design with a very long inlet trench caused problems with water distribution and, therefore, the inlet trench was subdivided into two or more separate units that could be loaded separately in order to better control water distribution (Brix, 1998). In the early 1980s, few HF CWs were built in other European countries such as Austria or Switzerland (Vymazal *et al.*, 1998).

In 1985, following visits to existing German and Danish systems, the first two HF CWs were built in the United Kingdom (here called Reed Bed Treatment Systems) and by the end of 1986, more than 20 HF CWs were designed (Cooper and Boon, 1987). The major change in the design was the use of very coarse filtration material, which ensured sub-surface flow similar to the former design by Seidel in 1965. Also, specific area of 5 m^2 PE^{-1} was used in the United Kingdom. In late 1980s, the first HF CWs were built in many European countries and HF CWs became the most widely used constructed wetland design in Europe by 1990s. The exchange of design and operational experience among nine European countries resulted in the European Design and Operations Guidelines for Reed Bed Treatment Systems (Cooper, 1990).

Typically, HF CWs (Figure 2) has a filtration bed depth of 0.6 to 0.8 m in order to allow roots of wetland plants and namely *Phragmites* to penetrate the whole bed and ensure bed oxygenation by oxygen release from roots. Roots and rhizomes of reeds and all other wetland plants are hollow and contain air-filled channels that are connected to the atmosphere for the purpose of transporting oxygen to the root system. The roots and rhizomes themselves for respiration use the majority of this oxygen, but as the roots are not completely gas-tight, some oxygen is lost to the rhizosphere (Brix, 1994). According to the working principle of HF CWs, the amount of oxygen released from roots and rhizomes should be sufficient to meet the demand for aerobic degradation of oxygen consuming substances in the wastewater, as well as for nitrification of ammonia. However, many studies have shown that the oxygen release from roots of different macrophytes is far less than the amount needed for aerobic degradation of oxygen consuming substances in sewage (Brix, 1990; Brix and Schierup, 1990). Despite that, the HF systems are very effective in removing organics and suspended solids (Table 2) and usually fulfil effluent discharge quality criteria, which for small sources of pollution are restricted to BOD_5 and suspended solids (Brix, 1994; Vymazal *et al.*, 1998). Removal of nutrients by these systems is low mainly because of the inability to oxidise ammonia, the predominant form of nitrogen in domestic and municipal sewage, and low sorption capacity of filtration media for phosphorus (Vymazal, 1999). Germany has the most HF CWs in operation; they exceed 50,000 (Wissing, pers.comm.). Other countries that commonly use HF systems are Austria (ca. 1000), United Kingdom (ca. 800), Denmark (ca 400), Italy (ca. 300), Czech Republic (ca 160), Poland, Portugal (ca. 120), Slovenia, France, Estonia, Norway and Switzerland.

Constructed wetlands with vertical sub-surface flow

Constructed wetlands with vertical flow were originally designed by Seidel (1965) as pre-treatment units before horizontal treatment. However, Seidel did not use VF wetlands as a single stage. The earliest full-scale system of this type were so called infiltration (or percolation) fields with vertical flow through a soil or sand medium and waters were discharged from the system via drains. This design was used to treat the wastewater from recreation sites

Table 2. Treatment efficiency of vegetated beds of HF CWs - world wide experience (Data from Australia, Austria, Brazil, Canada, Czech Republic, Denmark, Germany, India, Mexico, New Zealand, Poland, Slovenia, Sweden, USA and UK).

Parameter	Inflow (mg l^{-1})	Outflow (mg l^{-1})	Efficiency (%)	Number of observations
BOD$_5$	108	16.0	82.5	164
COD	284	72	74.6	131
TSS	107	18.1	83.1	158
TP	8.74	5.15	41.1	149
TN	46.6	26.9	42.3	137
NH$_4^+$-N	38.9	20.1	48.3	151
NO$_3^-$-N	4.38	2.87	34.5	79

outside the municipal network in Lauwersoog in the Netherlands in 1975 (Greiner and de Jong, 1984, Butijn and Greiner, 1985). The system consisted of a preliminary settling/distribution ditch, four infiltration compartments and an effluent ditch (Figure 3). The infiltration ditches had a net surface area of 1.3 ha and were planted with *Phragmites australis*. The drainage system was 0.55 m beneath the ground level. Raw domestic sewage was discharged into the preliminary settling/distribution ditches. After settling, water was fed every three to four days into one of the compartments. Because of the low loading rates and the fact that there were four ponds, each pond was dry for a period of 10 to 11 days (Greiner and de Jong, 1984). Treatment performance was high (Butijn and Greiner, 1985); however Veenstra (1998) indicates that due to strict guidelines imposed by Dutch Water Boards for conventional secondary treatment plants for < 20,000 PE, the use of single VF wetlands for large scale domestic sewage treatment is generally not considered in the Netherlands. Vertical flow systems are used to treat wastewater from dairy farms in the Netherlands (Veenstra, 1998).

At present, VF CWs usually comprise of a flat bed of graded gravel topped with sand and planted with macrophytes. The size fraction of gravel is larger in the bottom (ca. 30-60 mm) and smaller in the top layer (ca. 6 mm). Vertical flow CWs are fed intermittently with a large batch of water, subsequently surface flooding occurs. Wastewater then gradually percolates down through the bed and it allows air to refill the bed. This kind of feeding provides good oxygen transfer and hence the ability to nitrify (Cooper *et al.*, 1996). Vertical flow CWs provide a good removal of organics and suspended solids; however denitrification processes, are limited, as ammonia-N is usually only converted to nitrate-N (Table 3). As a consequence, removal of total nitrogen is low, usually lower than in HF CWs. Removal of phosphorus is low unless special filtration material with high sorption capacity is used (Mæhlum and Jenssen, 1998). Media such as Leca (light expanded clay aggregates) has been used in systems located in Norway, Estonia and Portugal. Vertical flow CWs require less land (1-2 m^2 PE^{-1}) as compared to HF systems (5-10 m^2 PE^{-1}), but require more maintenance and operation because of pumps, timers and other electric and mechanical devices required. Vertical flow systems are very popular in Austria, France and Germany, especially for small sources of pollution. In Denmark, a two-stage compact VF system has been designed, recently (Johansen *et al.*, 2002).

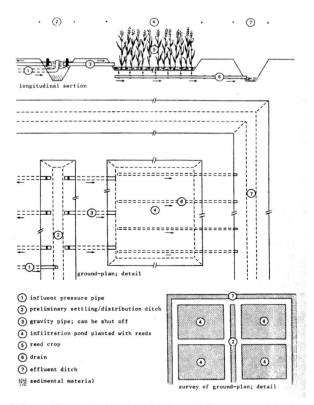

longitudinal section

ground-plan; detail

① influent pressure pipe
② preliminary settling/distribution ditch
③ gravity pipe; can be shut off
④ infiltration pond planted with reeds
⑤ reed crop
⑥ drain
⑦ effluent ditch
▓ sedimental material

survey of ground-plan; detail

Figure 3. Schematic ground plan and longitudinal section of a planted infiltration pond (Greiner and de Jong, 1984).

Table 3. Treatment efficiency of VF CWs (data from Australia, Austria, China, Denmark, France, Germany, Ireland, Poland, Nepal, Norway, Turkey and United Kingdom).

Parameter	Inflow (mg l^{-1})	Outflow (mg l^{-1})	Efficiency (%)	Number of observations
BOD$_5$	145	27.2	81.6	25
COD	303	75	71.3	16
TSS	97	18.4	77.5	14
TP	8.6	4.4	52.6	16
TN	61	35	39.7	18
NH$_4^+$-N	45.6	15.8	65.1	22
NO$_3^-$-N	1.55	15.1		16

Hybrid constructed wetlands

Various types of CWs may be combined in order to achieve higher treatment effectiveness, especially for nitrogen. Hybrid systems comprise most frequently of VF and HF systems arranged in a staged manner. Tertiary treatment of wastewaters by HF systems produce well-nitrified effluents (Cooper *et al.*, 1996); however secondary treatment of wastewaters by HF systems cannot do this because of the limited availability of oxygen in the system. Vertical flow systems do provide good conditions for nitrification as oxygen is available, but no denitrification can occur because of the presence of oxygen. The advantages and disadvantages of the HF and VF systems can be combined to complement each other. It is possible to produce an effluent low in O_2 demanding organics (BOD), which is fully nitrified and partly denitrified and hence has much lower total-N content (Cooper, 1999, 2001).

Many of these hybrid systems are derived from the original hybrid systems developed by Seidel at the Max Planck Institute in Krefeld, Germany. The process (Figure 4) is known as the Seidel system, the Krefeld system or the Max Planck Institute Process (MPIP) (Seidel, 1965, 1976, 1978). The design consists of two stages of several parallel VF beds followed by two or three HF beds in series. The VF stages are usually planted with *Phragmites australis*, whereas the HF stages contain a number of other emergent macrophytes, including *Iris, Schoenoplectus, Sparganium, Carex, Typha* or *Acorus*. The VF beds are loaded with pre-treated wastewater for one to two days, and were then allowed to dry out for four to eight days. The thin crust of

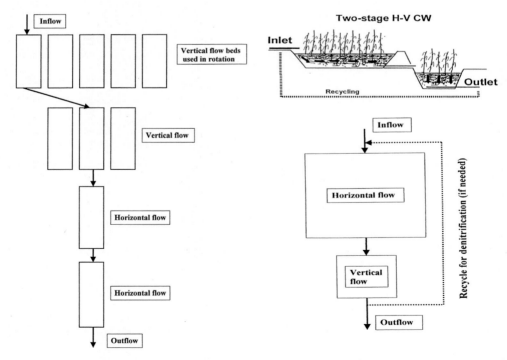

Figure 4. Hybrid systems based on concepts by Seidel (left) and Brix and Johansen (right) (Brix, 1998; Cooper, 1999).

solids that forms on top of the VF beds is mineralized during the rest period and achieves an equilibrium thickness (Brix, 1994).

In the early 1980s, several hybrid systems of Seidel´s type were built in France such as the one at Saint Bohaire, which was put into operation in 1982 (Boutin, 1987; Vuillot and Boutin, 1987). It consisted of four and two parallel VF beds in the first and second stages respectively, the third, fourth and fifth stages consisted of one HF bed in each stage. A similar system was built in 1987 in the UK at Oaklands Park (Burka and Lawrence, 1990). In 1990s and early 2000s, VF-HF systems were built in many European countries, e.g. in Slovenia (Urbanc-Bercic and Bulc, 1994), Norway (Mæhlum and Stalnacke, 1999), Austria (Mitterer-Reichmann, 2002), France (Reeb and Werckmann, 2003) and Ireland (O´Hogain, 2003) and now this type is getting more attention in most European countries. In table 4 typical treatment effectiveness of a VF-HF system is presented. The Colecott system consists of four VF beds (total of 64 m^2) at the first stage, 2 VF beds (60 m^2) at the second stage and one HF bed (60 m^2) at the third stage and is designed for 60 PE.

In the mid-1990s, Johansen and Brix (1996) introduced a HF-VF hybrid system (Figure 4). The large HF bed is placed first to remove organics and suspended solids and to provide denitrification. An intermittently loaded small VF bed is designed for further removal of organics and suspended solids and to nitrify ammonia to nitrate. However, in order to remove total N, the nitrified effluent from the VF bed must be recycled to the sedimentation tank. Brix *et al.* (2003) pointed out that special care must be taken not to affect the performance of the sedimentation tank or the nitrifying capacity of the VF bed by recycling too large volumes of wastewater. A similar system was built in Poland at Sobiechy (Ciupa, 1996).

Recently, hybrid CWs often includes FWS systems. A CW system that treats municipal wastewater at Kõo, Estonia consists of two VF beds (each 64 m^2 planted with *Phragmites australis*), followed by HF bed (350 m^2 planted with *Phragmites australis* and *Typha latifolia*) and two FWS wetlands (3600 m^2 and 5000 m^2 planted with *Typha latifolia*). The removal of BOD, total nitrogen and total phosphorus amounted to 88%, 65% and 72% (Mander *et al.*, 2003). In Italy, hybrid CWs are successfully used for treatment of concentrated winery wastewaters (Masi *et al.*, 2002). System at Ornellaia consists of two VF beds (90 m^2 each), followed by HF bed (102 m^2) and FWS wetland (148 m^2) and the pond (338 m^2). A system at Cecchi consists of HF bed (480 m^2) followed by FWS wetland (850 m^2). The organic load of the HF bed at Cecchi amounted to 1336 kg ha^{-1} d^{-1}. Treatment effects were high for organics, suspended solids and nutrients (Table 5, Masi *et al.*, 2002).

Sludge drying reed beds

In the last 35 years, interest has developed in using reeds to enhance the performance of conventional sludge drying beds. The first attempts were carried out in Karlsruhe, Germany in the late 1960s (Bittman and Seidel, 1967; Kickuth, 1969). The process was then used at other German sewage works sites in the 1970s and 1980s (Lienard *et al.*, 1990). In late 1980s and 1990s sludge drying beds were introduced to many European countries (France, U.K., Poland, and first of all to Denmark where, at present, about 25% of all sludges from conventional sewage plants are mineralised using reed beds). In Denmark, about 105 systems are in

Table 4. Performance of the VF-HF Colecott hybrid system (data from O 'Hogain, 2002). Concentrations are in mg l^{-1} and removal is in percent reduction. VF1out is outflow of VF bed one; VF2out is outflow from VF bed two and HFout is outflow from a final HF bed.

Parameter	Inflow	VF1out	VF2out	HFout	Removal
COD	462	210	66	47	89.4
BOD_5	269	171	43	23	91.4
TSS	53	28	3	1	98.1
NH_4^+-N	45	28	16	7	84.4
NO_3^--N	0.1	4.7	3.8	2.7	
NO_2^--N	0.1	0.2	0.1	0.1	
PO_4	18	16	15	11	38.9

Table 5. Casa Vincicola Cecchi, Italy - treatment performance for the period of February 2001 until March 2003 (based on Masi et al., 2002).

	BOD_5	COD	TSS	TN	TP	pH
Inflow	1833	3906	213	18.9	4.7	6.1
HF out	49.4	131	13.3	4.8	1.5	6.9
FWS out	25.4	84	23.4	3.5	1.3	7.4

operation with the largest systems in Skive and Kolding serving 123,000 and 125,000 PE (Nielsen, 2003). The Danish experience indicates that reed basins for surplus activated and digested sludge must be dimensioned to loading rates of maximum 60 and 50 kg dry matter m^{-2} yr^{-1} and minimum of 8 and 10-12 basins, respectively. Current types of drying beds are capable of treating various types of sludge with dry matter content of approximately 0.5 to 5%. After 10 years of drying the dry matter content in the sludge residue is about 40% (Nielsen, 2003).

Types of wastewater

Constructed wetlands have traditionally been used in Europe for treatment of domestic and municipal sewage from both separate and combined sewerage (Vymazal et al., 1998). However, since late 1980s CWs have been used for many other types of wastewater, including agricultural wastewaters (cattle, swine, poultry, dairy), mine drainage, food processing wastewaters (winery, abattoir, fish, potato, vegetable, meat, cheese, milk, sugar production), heavy industry wastewaters (polymers, fertilizers, chemicals, oil refineries, pulp and paper mills), landfill leachate and runoff waters (urban, highway, airport, nursery, greenhouse) (Cooper and Findlater, 1990; Vymazal et al., 1998; Nehring and Brauning, 2002; Dias and Vymazal, 2003; Mander and Jenssen, 2003).

Acknowledgements

The study was supported by grant MSM 000020001 "Solar Energetics of Natural and Technological Systems" from the Ministry of Education and Youth of the Czech Republic and by grant No. 206/02/1036 "Processes Determining Mass Balance in Overloaded Wetlands" from the Grant Agency of the Czech Republic.

References

Bittman, M. and K. Seidel, 1967. Entwässerung und Aufbereitung von Chemieschlamm mit Hilfe von Pflanzen. GWF **108** 488-491.

Boon, G.A., 1986. Report of a visit by members and staff of WRc to Germany (GFR) to investigate the Root Zone Method for treatment of waste waters. WRc Report 376-S/1, Swindon, UK, 61 pp.

Börner, T., K. von Felde, T. Gschlössl, E. Gschlössl, S. Kunst, and F.W. Wissing, 1996. Germany. In: Constructed Wetlands for Wastewater Treatment, edited by J. Vymazal, H. Brix, P.F. Cooper, B. Green and R. Haberl, Backhuys Publishers, Leiden, The Netherlands, pp. 169-190.

Boutin, C., 1987. Domestic wastewater treatment in tanks planted with rooted macrophytes: case study; description of the system; design criteria; and efficiency. Water Science and Technology **19**, 29-40.

Brix, H., 1987. Treatment of wastewater in the rhizosphere of wetland plants - the root-zone method. Water Science Technology **19** 107-118.

Brix, H., 1990. Gas exchange through the soil-atmosphere interphase and through dead culms of *Phragmites australis* in a constructed reed bed receiving domestic sewage. Water Research **24** 259-266.

Brix, H., 1994. Constructed wetlands for municipal wastewater treatment in Europe. In: Global Wetlands: Old World and New, edited by W.J. Mitsch, Elsevier Science B.V., Amsterdam, The Netherlands, pp. 325-333.

Brix, H., 1998. Denmark. In: Constructed Wetlands for Wastewater Treatment in Europe, edited by J. Vymazal, H. Brix, P.F. Cooper, B. Green and R. Haberl, Backhuys Publishers, Leiden, The Netherlands, pp. 123-152.

Brix, H. and H.H. Schierup, 1989. The use of macrophytes in water pollution control. Ambio **18** 100-107.

Brix, H. and H.H. Schierup, 1990. Soil oxygenation in constructed reed beds: the role of macrophyte and soil-atmosphere interface oxygen transport. In: Constructed Wetlands in Water Pollution Control, edited by P.F. Cooper and B.C. Findlater, Pergamon Press, Oxford, UK, pp. 53-66.

Brix, H., C. Arias, and N.H. Johansen, 2003. Experiments in a two-stage constructed wetland system: nitrification capacity and effects of recycling on nitrogen removal. In: Wetlands: Nutrients, Metals and Mass Cycling, edited by Jan Vymazal, Backhuys Publishers, Leiden, The Netherlands, pp. 237-258.

Burka, U. and P. Lawrence, 1990. A new community approach to waste treatment with higher water plants. In: Constructed Wetlands in Water pollution Control, edited by P.F. Cooper and B.C. Findlater, Pergamon Press, Oxford, U.K., pp. 359-371.

Butijn, G.D. and R.W. Greiner, 1985. Afwalwaterzuivering met behulp van begroeide infiltratievelden (Wastewater treatment with vegetated percolation fields). In: Wetlands for the Purification of Wastewater, edited by P.J.M. van der Aart, Department of Plant Ecology, University of Utrecht, The Netherlands, pp. 64-89.

Ciupa, R., 1996. The experience in the operation of constructed wetlands in North-Eastern Poland. In: Proc. 5th Internat. Conf. Wetland Systems for Water Pollution Control, IWA and Universität für Bodenkultur, Vienna, Chapter IX/6.

Cooper, P.F. (editor), 1990. European Design and Operations Guidelines for Reed Bed Treatment Systems. Prepared for the European Community/European Water Pollution Control Association Emergent Hydrophyte Treatment System Expert Contact Group. WRc Report UI 17, Swindon, UK.

Cooper, P.F., 1999. A review of the design and performance of vertical flow and hybrid reed bed treatment systems. Water Science and Technoloygy **40**(3) 1-9.

Cooper, P.F., 2001. Nitrification and denitrification in hybrid constructed wetlands systems. In: Transformations on Nutrients in Natural and Constructed Wetlands, edited by J. Vymazal, Backhuys Publishers, Leiden, The Netherlands, pp. 257-270.

Cooper, P.F. and A.G. Boon, 1987. The use of *Phragmites* for wastewater treatment by the Root Zone Method: The UK approach. In: Aquatic Plants for Water Treatment and Resource Recovery, edited by K.R. Reddy and W.H. Smith, Magnolia Publishing, Orlando, Florida, pp. 153-174.

Cooper, P.F. and B.C. Findlater (editors), 1990. Constructed Wetlands in Water Pollution Control. Pergamon Press, Oxford, UK, pp. 605.

Cooper, P.F., G.D. Job, M.B. Green, and R.B.E. Shutes, 1996. Reed Beds and Constructed Wetlands for Wastewater Treatment. WRc Publications, Medmenham, Marlow, UK, pp. 184.

Czerwenka, K. and K. Seidel, 1965. Neue Wege zur Grundwasseranreicherung. Das Gas- und Wasserfach **30** 828-831.

De Jong, J., 1976.. The purification of wastewater with the aid of rush or reed ponds. In: Biological Control of Water Pollution, edited by J. Tourbier and R.W. Pierson, Pennsylvania University Press, Philadelphia, pp. 133-139.

Dias, V. and J. Vymazal (editors), 2003. The Use of Aquatic Macrophytes for Wastewater Treatment in Construcvted Wetlands. Instituto da Conservação da Narurreza ans Instituto Nacional da Água, Lisbon, Portugal, 647 pp.

Greiner, R.W. and J. de Jong, 1984. The Use of Marsh Plants for the Treatment of Waste Water in Areas Designated for Recreation and Tourism. RIJP Report No. 225, Lelystad, The Netherlands, 33 pp.

Johansen, N.H. and H. Brix, 1996. Design criteria for a two-stage constructed wetland. In: Proc. 5th Internat. Conf. Wetland Systems for Water Pollution Control, IWA and Universität für Bodenkultur, Vienna, Chapter IX/3.

Johansen, N.H., H. Brix, and C. Arias, 2002. Design and characterization of a compact constructed wetland system removing BOD, nitrogen and phosphorus from single household sewage. In: Proc. 8th Internat. Conf. Wetland Systems for Water Pollution Control, IWA and University of Dar es Salaam, pp. 47-61.

Kickuth, R., 1969. Höhere Wasserpflanzen und Gawässerreinhaltung. Schiftenreihe der Vereinigung Deutscher Gewässerschutz EV-VDG **19** 3-14.

Kickuth, R., 1977. Degradation and incorporation of nutrients from rural wastewaters by plant rhizosphere under limnic conditions. In: Utilization of Manure by Land Spreading, Comm. Europ. Commun., EUR 5672e, London, UK, pp. 335-343.

Kickuth, R., 1978. Elimination gelöster Laststoffe durch Röhrichtbestände. Arbeiten des Deutsches Fischereiverbandes **25** 57-70.

Kickuth, R., 1981. Abwasserreinigung in Mosaikmatritzen aus aeroben und anaeroben Teilbezirken. In: Grundlagen der Abwasserreinigung, edited by F. Moser, Verlag Oldenburg, München, Wien, pp. 630-65.

Kowalik, P. and H. Obarska-Pempkowiak, 1998. Poland. In: Constructed Wetlands for Wastewater Treatment in Europe, edited by J. Vymazal, H. Brix, P.F. Cooper, B. Green and R. Haberl, Backhuys Publishers, Leiden, The Netherlands, pp. 217-225.

Lakatos, G., 1998. Hungary. In: Constructed Wetlands for Wastewater Treatment in Europe, edited by J. Vymazal, H. Brix, P.F. Cooper, B. Green and R. Haberl, Backhuys Publishers, Leiden, The Netherlands, pp. 191-206.

Lakatos, G., M.K. Kiss, and P. Juhász., 1996. Application of constructed wetlands for wastewater treatment in Hungary. In: Proc. 5th Internat. Conf. Wetland Systems for Water Pollution Control, IWA and Universität für Bodenkultur, Vienna, Chapter IX/9.

Lienard, A., D. Esser, A. Dequin, and F. Virloget, 1990. Sludge dewatering and drying beds: an interesting solution? General investigations and first trials in France. In: Constructed Wetlands in Water Pollution Control, edited by P.F. Cooper and B.C. Findlater, Pergamon Press, Oxford, UK, pp. 257-267.

Mander, Ü., S. Teiter, K. Lõhmus, T. Mauring, K. Nurk, and J. Augustin, 2003. Emission rates of N_2O and CH_4 in riparian alder forest and subsurface flow constructed wetland. In: Wetlands: Nutrients, Metals and Mass Cycling, edited by Jan Vymazal, Backhuys Publishers, Leiden, The Netherlands, pp. 259-279.

Mander, Ü. and P. Jenssen (editors), 2003. Constructed Wetlands for Wastewater Treatment on Cold Climates. WIT Press, Southampton, UK, pp. 325.

Masi, F., G. Conte, N. Martinuzzi, and B. Pucci, 2002. Winery high organic content wastewaters treated by constructed wetlands in Mediterranean climate. In: Proc. 8th Internat. Conf. Wetland Systems for Water Pollution Control, IWA and University of Dar es Salaam, pp. 274-282.

Mæhlum, T. and P. Jenssen, 1998. Norway. In: Constructed Wetlands for Wastewater Treatment in Europe, edited by J. Vymazal, H. Brix, P.F. Cooper, B. Green and R. Haberl, Backhuys Publishers, Leiden, The Netherlands, pp. 207-216.

Mæhlum, T. and P. Stalnacke, 1999. Removal efficiency of three cold-climate constructed wetlands treating domestic wastewater: effects of temperature, seasons, loading rates and input concentrations. Water Science and Technology **40** (3) 273-281.

Mitterer-Reichmann, G.M., 2002. Data evaluation of constructed wetlands for treatment of domestic wastewater. In: Proc. 8th Internat. Conf. Wetland Systems for Water Pollution Control, IWA and University of Dar es Salaam, pp. 40-46.

Nehring, K.W. and S.E. Brauning (editors), 2002. Wetlands and Remediatrion II. Battelle Memorial Institute, Columbus, Ohio, 386 pp.

O´Hogain, S., 2003. The design, operation and performance of a municipal hybrid reed bed treatment system. Water Science and Technology **48**(5) 119-126.

Reeb, G. and M. Werckmann, 2003. Looking at the outlet zone of three constructed wetlands treating wastewaters of small communities. In: Wetlands - Nutrientrs, Metals and Msss Cycling, edited by J. Vymazal, Backhuys Publishers, Leiden, The Netherlands, pp. 191-199.

Seidel, K., 1955. Die Flechtbinse *Scirpus lacustris*. In: Ökologie, Morphologie und Entwicklung, ihre Stellung bei den Volkern und ihre wirtschaftliche Bedeutung, Schweizerbart´sche Verlagsbuchnadlung, Stuttgart, Germany, pp. 37-52..

Seidel, K., 1961. Zur Problematik der Keim- und Pflanzengewasser. Verh. Internat. Verein. Limnol. **14** 1035-1039.

Seidel, K., 1965 a. Phenol-Abbau in Wasser durch *Scirpus lacustris* L. wehrend einer versuchsdauer von 31 Monaten. Naturwissenschaften **52** 398-406.

Seidel, K., 1965 b. Neue Wege zur Grundwasseranreicherung in Krefeld. Vol. II. Hydrobotanische Reinigungsmehode. GWF Wasser/Abwasser **30** 831-833.

Seidel, K., 1966. Reinigung von Gewässern durch höhere Pflanzen. Naturwissenschaften **53** 289-297.

Seidel, K., 1976. Macrophytes and water purification. In: Biological Control of Water Pollution, edited by J. Tourbier and R.W. Pierson, Pennsylvania University Press, Philadelphia, pp. 109-122.

Seidel, K., 1978. Gewässerreinigung durch höhere Pflanzen. Zeitschrift Garten und Landschaft **H1** 9-17.

Strusevicius, Z. and S.M. Struseviciene, 2003. Investigations of wastewater produced on cattle-breeding farms and its treatment in constructed wetlands. In: Proc. Conf. Constructed and Riverine Wetlands for Optimal Control of Wastewater at Catchment Scale, edited by U. Mander, C. Vohla and A. Poom, University of Tatru, Estonia, pp. 317-423.

Sundblad, K., 1998. Sweden. In: Constructed Wetlands for Wastewater Treatment in Europe, edited by J. Vymazal, H. Brix, P.F. Cooper, B. Green and R. Haberl, Backhuys Publishers, Leiden, The Netherlands, pp. 251-259.

Urbanc-Bercic, O. and T. Bulc, 1994. Integrated constructed wetland for small communities. In: Proc. 4th Internat. Conf. Wetland Systems for Water Pollution Control, ICWS´94 Secretariat, Guangzhou, P.R. China, pp. 138-146.

Veenstra, S., 1998. The Netherlands. In: Constructed Wetlands for Wastewater Treatment in Europe, edited by J. Vymazal, H. Brix, P.F. Cooper, B. Green and R. Haberl, Backhuys Publishers, Leiden, The Netherlands, pp. 289-314.

Vuillot, M. and C. Boutin, 1987. Les systèmes rustiques d´épuration: aspects de l´expérience française; possibilités d´application aux pays en voie de dévelopment. Trib. Cebedeau 518 **40** 21-31.

Vymazal, J., 1999. Nitrogen removal in constructed wetlands with horizontal sub-surface flow-can we determine the key process? In: Nutrient Cycling and Retention in Natural and Constructed Wetlands, edited by J. Vymazal, Backhuys Publishers, Leiden, The Netherlands, pp. 1-17.

Vymazal, J., 2001. Types of constructed wetlands for wastewater treatment: their potential for nutrient removal. In: Transformations on Nutrients in Natural and Constructed Wetlands, edited by J. Vymazal, Backhuys Publishers, Leiden, The Netherlands, pp. 1-93.

Vymazal, J., H. Brix, P.F. Cooper, M. Green and R. Haberl, (editors), 1998. Constructed Wetlands for Wastewater Treatment in Europe. Backhuys Publishers, Leiden, The Netherlands, pp. 348.

Microbiological studies on an integrated constructed wetland used for treatment of agricultural wastewaters

S. McHugh[1], K. Richards[2], E.J. Dunne[2,3], R. Harrington[4] and V. O'Flaherty[1]
[1]Microbial Ecology Laboratory, Department of Microbiology, National University of Ireland, Galway, Ireland
[2]Teagasc Research Centre, Johnstown Castle, Wexford, Ireland
[3]Wetland Biogeochemistry Laboratory, Soil and Water Science Department, University of Florida/IFAS, 106 Newell Hall, PO Box 110510, Gainesville, FL 32611-0510, USA
[4]National Parks and Wildlife, Department of Environment, Heritage and Local Government, The Quay, Co. Waterford, Ireland

Abstract

Culture-independent, molecular techniques were used to investigate the microbial community structure of an integrated constructed wetland (ICW) used for the treatment of agricultural wastewater. Procedures were optimised for total DNA recovery and polymerase chain reaction (PCR) amplification of 16S rRNA genes using archaea- and bacteria-specific oligonucleotide primers. Cloned PCR products were subsequently screened by amplified rDNA restriction analysis to identify operational taxonomic units (OTUs). Inserts from clones representing selected OTUs were sequenced and phylogenetic trees were prepared. A high level of bacterial diversity was observed within the wetland, with the main species detected belonging to the *Proteobacteria*, the Gram Positive Bacteria and the *Bacteroides-Flexibacter-Cytophaga* group. Sequences closely related to organisms involved in the nitrogen (*Nitrosomonas*, *Nitrospina* sp.) and carbon (*Methanosaeta* sp., *Methylobacter* sp.) cycles were also detected, indicating the important role of microbes in nutrient cycling within wetland ecosystems.

Keywords: wetlands, 16S rRNA gene, bacteria, clone libraries, greenhouse gas.

Introduction

Constructed or artificial wetlands are currently recognised as an effective treatment option for a wide range of wastewaters and effluents. Although these treatment systems are largely dependant on the actions and interactions of microbial communities, research into the microbial ecology of wetland environments is limited. Microbial transformations are reported to be responsible for the bulk of water quality improvement in wetland treatment systems, with micro-organisms playing key roles in the degradation of organic matter, including hazardous and recalcitrant compounds, nutrient (nitrogen and phosphorus) removal and metal transformations. The diversity of microbes in a wetland environment, therefore, may be critical for the functioning and maintenance of the system. Furthermore, anaerobic microbial activity is likely to result in the release of N_2O and CH_4 to the atmosphere and so the microbial populations present may be directly linked to greenhouse gas emissions. An increased insight into the microbial ecology of wetland environments is essential to the further development and optimisation of this treatment technology.

In the present study, the microbial community structure of an Integrated Constructed Wetland (ICW) used to treat agricultural wastewater was investigated using clone library analysis, amplified rDNA restriction analysis and 16S rRNA gene sequencing. These nucleic acid-based molecular techniques allow direct identification of micro-organisms based on their DNA sequence and, thus, eliminate many of the problems associated with traditional microbiological methods (Amann *et al.*, 1995).

Materials and methods

Sediment samples were taken from an ICW located at the organic dairy farm in the Teagasc Research Centre, Johnstown Castle, Co. Wexford, Ireland. The wetland treated farmyard dirty water from a 42 cow unit (open yard area of 2,030 m^2) and had an overall area of 4,800 m^2 that comprised of three treatment cells and one final monitoring pond. Operational and chemical data produced on the performance of the farm-scale ICW showed significant decreases in mass loads of phosphate (PO$_4$-P), ammonia (NH$_4$-N), five-day biochemical oxygen demand (BOD$_5$) and total suspended solids (TSS). DNA was extracted, in triplicate, from biomass sampled from the three treatment cells using a chemical lysis extraction method (McHugh *et al.*, 2003). One archaeal (from treatment wetland cell 1) and three bacterial (from wetland cells 1, 2 and 3) clone libraries were generated by PCR amplification of 16S rRNA genes using the archaeal primers 21F (5' - TTCCGGTTGATCCYGCCGGA - 3') and 958R (5' - YCCGGCGTTGAMTCCAATT - 3') and the bacterial primers 27F (5' - GAGTTTGATCCTGGCTCAG - 3') and 1392R (5' - ACGGGCGGTGTGTRC - 3'), as described previously (McHugh *et al.*, 2004). Insert-containing clones were screened using amplified rDNA restriction analysis (ARDRA) and grouped into OTUs by visualisation on a 3.5% high resolution agarose gel (McHugh *et al.*, 2003). Representatives from selected OTUs were sequenced and the resultant sequence data was compared to nucleotide databases using the basic local alignment search tool (BLASTn; Altschul *et al.*, 1997) and manually aligned to sequences obtained from the Ribosomal Database Project (RDP; Maidak *et al.*, 1999). The phylogenetic inference package Paup* 4.0b8 was used for all phylogenetic analysis (Swofford, 2001).

Results and discussion

Sequence data revealed that 78% of the archaeal clones analysed from treatment wetland cell 1 (ARC1, ARC3, ARC4, ARC6; Figure 1) were closely related to *Methanosaeta* sp. This acetoclastic methanogen is often prevalent in soil and sediment samples and converts acetate to methane (CH$_4$) in anaerobic environments. Eleven percent of the clones analysed were affiliated to the hydrogenotrophic *Methanomicrobiales* group (ARC2, ARC7; Figure 1), which convert hydrogen and carbon dioxide to CH$_4$. The presence of both acetoclastic and hydrogenotrophic methanogens would indicate significant CH$_4$ emissions from treatment wetland cell 1. Clones related to the *Crenarchaeota* were also detected in the sediment sample (ARC5; Figure 1). Metabolically, the *Crenarchaeota* are quite diverse, but several species are known to gain energy from the fermentation of organic substrates. The presence of these organisms in wetland sediment, and also in a number of other ecologically important temperate environments, such as fresh-water lake sediments, anaerobic sludges and soybean soil, implies that previous assumptions that the *Crenarchaeota* are exclusively extremophiles, and thus not broadly of ecological significance, may not be correct.

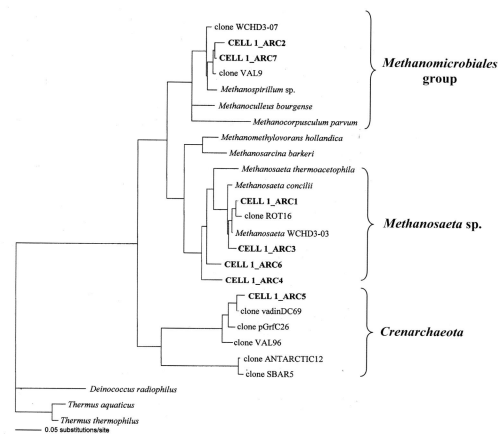

Figure 1. Phylogenetic tree constructed with evolutionary distances calculated based on the Kimura-2 model and the neighbour joining method of Saitou and Nei (1987). Three bacterial sequences were defined as outgroups during phylogenetic reconstruction.

Analysis of the bacterial community structure within treatment wetland cells 1, 2 and 3 reveal the bacterial populations to be considerably more diverse than the archaeal populations, with each clone analysed by ARDRA producing a unique banding pattern. Seventeen, 12 and 13 random representatives from wetland cells 1, 2 and 3, respectively, were sequenced. Members of the Gram Positive Bacteria, the *Flexibacter-Cytophaga-Bacteroides* group and the *Proteobacteria* (beta and delta sub-groups) were present in all three cells. Clones related to the *Prosthecobacter* group were present in wetland cells 1 and 2 only, while clones related to the WCHB1-31 group were detected in cells 2 and 3. *Spirochetes* were detected in wetland cell 1 and members of the *Nitrospina* and *Planctomyces* groups were found in cell 2.

The majority of retrieved sequences corresponded to microbes, which are capable of degrading organic matter and are, therefore, responsible for water quality improvement within the wetland. As specific microbial trophic groups are known to be responsible for the breakdown of certain organic compounds, knowledge of the microbial composition of a wetland may allow assessment of components of treatment rates of the system. Furthermore, population profiles may help to explain ecosystem productivity, performance and process failure.

Two sequences obtained from treatment wetland cell 1 were related to the *Nitrosomonas* group. These are chemolithotrophic ammonia oxidising bacteria, which convert NH_4 to nitrite (NO_2^-). One sequence obtained from wetland cell 2 was related to the *Nitrospina* sp., which are nitrite oxidising bacteria that convert NO_2^- to nitrate (NO_3^{2-}). Also in wetland cell 2, one retrieved sequence showed strong similarity to the methanotrophic organism *Methylobacter* sp. Methanotrophs (methane-oxidising bacteria) can consume a significant amount of the methane gas produced in a wetland and so play an important role in the reduction of greenhouse gas emissions. Many of the clones analysed in this study were closely related to sequences for which no function is yet assigned. Also, due to the high bacterial diversity present in each sample and the relatively small number of clones sequenced, it is likely that the presence of other notable micro-organisms (eg. dinitrifiers, phosphate-accumulating organisms) may have been overlooked. This study does, however, demonstrate that a more complete understanding of the processes occurring within a wetland environment can be obtained by examining the microbial communities present. Nevertheless, it is apparent that much fundamental research, using a broad range of techniques and experimental approaches, is required.

Conclusions

Through the use of modern microbiological techniques, a greater insight into the microbial ecology of constructed wetland ecosystems can be obtained. This is essential to the further development and optimisation of wetlands as a wastewater treatment technology, as microbial processes regulate the major nutrient cycles and degradation of organic matter occurring within these systems. Furthermore, knowledge of the identity and function of the microbial populations present in a wetland may allow the tracking of nitrogen and carbon, in the context of greenhouse gas emissions, which is an important consideration in using wetlands as a method to treat wastewaters.

Acknowledgements

This work was funded by the National University of Ireland, Galway Millennium Fund.

References

Altschul, S., T.L. Madden, A.A. Schaffer, J. Zhang., W. Miller, and D.J. Lipman, 1997. Gapped BLAST and PSI-BLAST: a new generation of protein database search programs. Nucleic Acids Research **25** 3389-3402.

Amann, R.I., 1995. Fluorescently-labelled, rRNA-targeted oligonucleotide probes in the study of microbial ecology. Molecular Ecology **4** 543-554.

Maidak, B.L., J.R. Cole, T.G. Lilburn, C.T. Parker Jr., P.R. Saxman, J.M. Stredwick, G.M. Garrity, B. Li, G.J. Olsen, S. Pramanik, T.M. Schmidt, and J.M. Tiedje, 1999. The RDP (Ribosomal Database Project) continues. Nucleic Acids Research **28** 173-174.

McHugh, S., M. Carton, T. Mahony, and V. O'Flaherty, 2003 Methanogenic population structure in a variety of anaerobic bioreactor sludges. FEMS Microbiolology Letters **219** 297-304.

McHugh, S., M. Carton, G. Collins, and V. O'Flaherty, 2004. Reactor performance and microbial community dynamics during anaerobic biological treatment of wastewaters at 16-37 °C, FEMS Microbiology Ecology **48** 369-378.

Saitou, N. and M. Nei, 1987. The neighbour-joining method: a new method of reconstructing phylogenetic trees. Molecular Biology and Evolution **4** 406-425.

Swofford, D.L., 2001. PAUP*. Phylogenetic Analysis Using Parsimony (*and Other Methods). Version 4. Sinauer Associates, Sunderland, Massachusetts.

Wetlands for the management of soiled water: a perspective from the Department of Agriculture and Food, Ireland

C. Robson
Specialist Farm Services, Department of Agriculture and Food, Kildare St., Dublin, Ireland

Responsibilities of the Department of Agriculture and Food

For thirty years, the Department of Agriculture and Food has administered grant-aided schemes, using mainly European Union (EU) funding, for Irish farm buildings and structures. These schemes are generous: the value added tax is reimbursed, and there are further grants to cover usually 40% of construction costs. As a quid-pro-quo, all structures must conform to one or more of the Department's minimum specifications. One of the responsibilities of the Department is to prepare these specifications, currently numbering about thirty, and continually to revise them in the light of new ideas, materials, and standards. Because many agricultural building firms are very small, with few engineering design skills, these specifications are effectively books on how to build. All the Department's specifications are available on www.agriculture.gov.ie.

The Department is also responsible for all other grant-aid to Irish farmers, which includes the Rural Environment Protection Scheme (REPS), area aid, cattle and sheep premia. The Department also has further direct involvement in the nitrates action programme and the implementation of "good farming practice." At every point, therefore, wetlands will impinge on our responsibilities, both financial and environmental.

A perspective on constructed wetlands

If constructed wetlands become, as seems likely, a fully recognised part of Irish agricultural practice, then it will be the job of the Department to ensure that the necessary specifications and protocols are integrated into the Department's system of minimum specified standards, particularly if they are to become part of our grant-aid schemes. This has led to the development of a Technical Working Group on constructed wetlands, earth-lined slurry stores and out-wintering animal pads. Others in the technical working group represent the Department of the Environment, Heritage, and Local Government; the Environmental Protection Agency; the Geological Survey of Ireland; the National Parks and Wildlife Service; and Teagasc. The points emphasised in this perspective are not from the working group, although in areas where the working group has already reached an agreement, I will make that clear.

From the Department's point of view, a constructed wetland is simply a system to deal with soiled or contaminated water emanating from the farmyard. The Department is aware that they can also have very considerable ecological and environmental advantages. The severe limits on what can go into these systems, however, must be both recognised (and publicised) as some farmers are already convinced that you can use wetlands to get rid of both animal slurries and a range of effluents. One of the agreed decisions of the working group has been that wetlands cannot even be used for the run-off effluent from out-wintering pads, as the organic material concentrations, as measured by five day biological oxygen demand (BOD_5) is anywhere from 6,000 to 11,000 mg l^{-1}. These concentrations are considered too high. It is now clear that farm wetlands will handle only two types of waste: dairy washings, and the contaminated rainwater that has fallen on the soiled areas of the farmyard. This is a relatively small proportion of all the wastes that farms produce.

One of the main themes of the Department's 30 years of work on farmyard structures has been the reduction, or where possible the elimination, of soiled water on every grant-aided farm. Firstly, it is specified that every new roofed building must have both gutters and down pipes to get the clean water away towards a suitable watercourse, without any possible contamination from yard wastes. Secondly, in our own work with farmers, and now with professional consultants, we use every method we can to minimise or eliminate dirty yards in favour of clean dry ones. Drains, channels, falls and ridges are introduced to ensure that contaminated rainwater cannot reach clean concrete, but goes instead straight to a collecting tank. Buildings are designed, or converted, so that animals are fed at all times either within their buildings, or are in adjacent slatted areas. A farmyard for beef or sheep, built to our specifications and guidelines, will produce effectively no soiled water, and there are thousands of examples that do just that. Grants are also regularly given for redesigns or extensions to take animals away from external feed standings, and feed them inside instead in dry very well ventilated buildings. It is both better management, and more profitable.

With regard to future policy; there are, of course, beef and suckler farms where animals are still fed outside, but if a well-run beef-farm produces effectively no soiled water, then a farmer may be better advised to spend money improving the farmyard set-up, rather than spend it constructing a wetland. Thus, wetlands do not seem intrinsically applicable to any type of enterprise other than a dairy farm. Dairy farms do inevitably produce washwaters and also produce soiled water from contaminated yards, (although even this can be minimised in a well-designed system). Properly constructed wetlands on dairy farms do indeed seem to have a potentially valuable role.

Management of soiled water

Management of soiled water involves money, and wetlands are relatively expensive. It is clear that even current contractual costs for completed wetlands amount to some €14,800 ha^{-1} for a wetland ranging in size from about 0.8 to 1.6 hectares. In a conversation about ten days ago with specialist researchers on this topic, this figure of €14,800 was described as "distinctly on the low side." Bear in mind that in the future, registered contractors will have to follow a specification that has still not been finalised, and that the costs of the 1.8 m high safety childproof fences for the first and final ponds are not included in the above figure. Nor have the new mandatory costs for site assessment and soil characterisation by qualified personnel; planning permission fees; initial discharge licence; in addition to annual costs of about €2,000 a year for water monitoring by the farmer and the relevant Local Authority. It is difficult to see how the cost from a registered contractor is likely to be less than about €25,000 for wetland that is about 1.2 hectares in size. I may be wrong, but probably not by a huge amount. That sum would allow for a very fair amount of restructuring work on the farmyard to eliminate most soiled water sources.

In addition to this, is the change in land use, i.e. the conversion of productive land to non-productive land (wetland). Not every farmer has a few convenient useless acres. When I discussed this point last month with my Belgian counterpart, he gave that contemptuous shrug that continentals do so well: "in Belgium, profitable production is expected from every square metre."

I have already alluded to the potential misuse of these systems by farmers who, either mistakenly or with intent, decide to fill them with slurries, and silage or other effluents, resulting in unacceptable discharges to our rivers. Once-a-year local authority checks do after all leave plenty of time for misuse! There is already an investigation underway in the Southwest of the country where there is a range of new wetlands, all installed without planning permission or indeed specific design standards, and where some systems seem to be used for the purposes just mentioned.

This has led to another agreed decision by the technical working group. If wetlands are introduced, then planning permission, site assessment and soil characterisation forms must all be included with a final signed "Completion Certificate" from a recognised contractor, before any wetland is accepted. It will then be in the right place, and expertly constructed. Misuse, it is hoped, may also be less likely.

There is still an important question which has not received much attention. What is to happen in fifteen or twenty year's time to all that phosphorus-laden rotting vegetation and nutrient-rich silt, when the farm wetland reaches the end of its useful existence? Twenty years is only half a professional life, and the Department's specialist unit has had to deal with collapsing slats, decaying metal roofs, and crumbling silos that were once thought, to be "good enough for 20 years." Our solutions to those problems have had to include new national standards, new specifications, and prohibitions on whole construction practices. There may indeed be a perfectly simple solution to defunct wetlands on smallish farms, but as yet it has not been elaborated.

When wetlands are being considered for the treatment of soiled water, we must, of course, set them against the current practice of applying soiled water to grassland areas at suitable (or sometimes unsuitable) times throughout the entire year. It has been thought that, provided the water was sprayed or applied over a wide and frequently changed area, that this gave acceptable results. Recent research has begun to question this practice, particularly with very large quantities of water, and this research has quite rightly been cited as a further argument in favour of wetlands. However, it is fair to point out that it also provides an equally strong argument for works to minimise or preferably eliminate soiled water, such as I have described. It is, after all, an entirely valid aim to retain all nutrient-rich wastes throughout the winter, and then apply them as fertilisers of known value at the correct time. At present the Department of Agriculture and Food is actively encouraging and grant-aiding aeration systems for slurry tanks that ensure such a supply of fertiliser is accessible at any appropriate time.

I have barely touched on the ecological value of wetlands, increase in biodiversity and environmental improvements that farm wetlands can bring. These seem substantial. I suppose what concerns us is that a farmer installing a wetland should be fully aware of just why he or she is being advised to do so. Clarity is essential: if the ecological improvements are one half of the benefits, and waste management the other, then the farmer must knowingly choose both halves.

Conclusion

In the overall context, the important thing is that constructed wetlands must clearly be seen as one alternative solution to the problem of contaminated water from Irish farmyards, rather than Irish farmyards being seen as servicing the provision of source material for constructed wetlands. Put it that way, it suddenly appears almost too obvious, and possibly even pointless to mention.

As a final word, it appears there are already real advantages in the use of properly constructed wetlands for Irish dairy farms, for Irish rivers, and the Irish environment as a whole. Responsibilities of the Department of Agriculture and Food include both ensuring that those advantages are made public, but also making clear that there are alternative approaches, and that what appear to be as-yet-unanswered questions, do require resolution.

Constructed wetlands for wastewater treatment in rural communities

M. Keegan and F. Clinton
Environmental Protection Agency (EPA), Johnstown Castle Estate, Wexford, Ireland

Abstract

The potential role of constructed wetlands (CWs) in treating wastewater from small communities and single house developments has been recognised in Ireland. Primary treated effluents may conceivably be further treated using constructed wetland systems. The Irish Environmental Protection Agency (EPA) published a guidance manual in 2000 setting out best practice for the treatment of sanitary waste from small population centres. The approaches suggested include the provision of primary wastewater treatment such as septic tanks followed by CWs for further treatment prior to discharge to groundwater or surface water. A detailed site assessment is outlined in the guidance note. It is recommended that the specified procedure should be followed in order to determine the suitability of a given site for the establishment of a constructed wetland. The site assessment includes: local planning restrictions, current legislation, hydrological and hydrogeological aspects, soil and subsoil, other activities in the locality and engineering challenges presented by the proposal to locate the constructed wetland. Issues related to the licensing of discharges from CWs are also addressed. Where discharges are to surface waters and where volumes in excess of $5m^3$/day are to be discharged to groundwater, a licence to discharge to waters is required under section 4 of the Water Pollution Acts 1977 and 1990.

Keywords: constructed wetlands, reed beds, small communities, single houses, site characterisation, site assessment.

Introduction

The EPA was established in 1992. Among its functions is the production of guidance for a range of activities, which present a potential to cause environmental pollution to air, water and land. Prior to 2000 there was very little guidance on what was required for single house wastewater treatments systems, other than standard regulation (SR6:1991). This regulation specifically dealt with the provision of septic tank systems with percolation to ground. Recognising that this guidance did not take sufficient account of a range of other options for the treatment of wastewater from single houses developments, the Agency published a guidance manual entitled "Wastewater treatment manuals: treatment systems for single houses" (EPA, 2000). In contrast to earlier guidance, the EPA guidance document placed the principle emphasis on site characterisation. The capacity of the existing soil and subsoil on site was assessed to determine its suitability for the further treatment and ultimate disposal to ground of partially treated wastewater.

A range of treatment options is presented in the guidance. Among these are septic tank systems (including a percolation area), filter systems and mechanical aeration systems both with polishing filters. Constructed wetlands are considered to be included in the group of systems generally described as filter systems.

Site characterisation

The purpose of site assessment is to determine the suitability of a site for an on-site treatment system. The assessment will also help to predict wastewater flow through subsoil and into subsurface material. For a subsoil to be effective as a medium for treating wastewater, it must retain the wastewater for a sufficient length of time, and it must be well aerated. Only after a site assessment has been completed can an on-site system be chosen. In designing a wastewater treatment system to treat and dispose of wastewater, two factors must be considered:
• the suitability of subsoil and groundwater conditions;
• the permissible hydraulic load on the subsoil.

To assess these factors a site characterisation is undertaken. This includes:
• a desk study;
• an on-site evaluation, consisting of;
  – a visual assessment;
  – a trial hole;
  – percolation tests.

The purposes of a desk study are to obtain information relevant to site suitability, to identify targets at risk and establish if there are site restrictions. A desk study involves the assessment of available data pertaining to the site and adjoining areas. Information collected from the desk study should include material related to the hydrological, hydrogeological and planning aspects of the site, which may be available in maps or reports. Hydrological aspects include locating the presence (if any) of streams, rivers, lakes, beaches, shellfish areas and/or wetlands while hydrogeological aspects include soil type, subsoil type, aquifer type and groundwater protection responses (GWPR)[1] for on-site systems for single houses. In addition, the location of any archaeological or natural heritage sites in the vicinity of the proposed site should be identified.

The on-site evaluation comprises a visual assessment of the site, the examination of a trial hole excavated in the soil and subsoil and the carrying out of percolation tests. The purposes of the visual assessment are to assess the potential suitability of the site, to assess potential targets at risk (e.g. adjacent wells) and to provide sufficient information to enable a decision-making on site suitability for the location, treatment and disposal of wastewater within the site.

The purposes of the trial hole are to determine the depth of the water table, the depth to bedrock and the soil and subsoil characteristics.

---

[1] Groundwater Protection Responses outline acceptable on-site wastewater treatment systems in each groundwater protection zone (DELG/EPA/GSI, 1999) and recommend conditions and/or investigations depending on the groundwater vulnerability, the value of the groundwater resource and the contaminant loading.

A percolation (permeability) test assesses the hydraulic assimilation capacity of the subsoil i.e., the length of time for the water level in the percolation hole to drop by a specified amount. The objective of the percolation test is to determine the ability of the subsoil to hydraulically transmit the treated effluent from the treatment system, through the subsoil to groundwater. The test also gives an indication of the likely residence time of the treated effluent in the upper subsoil layers and therefore it provides an indication of the ability of subsoil to treat the residual pollutants contained in the treated effluent.

Selecting a system

The information collected from the desk study and on-site assessment should be used in an integrated way to determine whether an on-site system can be employed as a favourable effluent treatment and disposal option. If so, the type of system that is appropriate and the optimal final disposal route for the treated wastewater is determined at this stage. Depending on the characteristics of the site, more than one option may be available.

A constructed wetland system is an option used to provide further treatment for pre-treated wastewater. A constructed wetland is an engineered shallow basin, which is sealed by either a natural[2] or synthetic liner. Constructed wetlands can be categorised into two major groups based on the wastewater flow path through the system. These are a horizontal flow (HF) systems and vertical flow systems.

In horizontal flow (HF) CWs, wastewater is loaded at one end of a flat to gently sloping bed of reeds and flows across the bed to the outfall end. If the surface of the wastewater is at or above the surface of the wetland media, the system is often called a free-water surface (FWS) horizontal-flow wetland. If the surface of the wastewater is below the surface of the wetland media, the system is called a sub-surface horizontal flow wetland. As the wastewater flows through the wetland, micro-organisms, which are attached to plant roots and support media, purify the wastewater. The support media can consist of soil, gravel and/or other suitable material.

In vertical-flow (VF) wetlands, the wastewater is intermittently distributed uniformly over the media, and gradually drains vertically to a drainage network at the base of the support media. As the wastewater drains vertically, air re-enters pore spaces in media. The media used in the vertical-flow wetland can consist of a layer of sand overlying a layer of gravel. Care should be taken to ensure that the incoming effluent is controlled, so as not to disturb or erode the upper sand layers in the wetland.

Where a constructed wetland is the selected option for wastewater treatment, it is important that the system is correctly designed and constructed, to provide sufficient treatment and hydraulic transfer of wastewater loads. When designing a wetland the following factors need to be taken into account:
- Number of persons served by the system (loading rate in m^3).
- Depth of the wetland should be 0.6 m.
- Slope of the base (i) of the wetland should be 0.5 - 2.0%.

---

[2] Subsoils with an in-situ or enhanced permeability of 1 x10^{-8} m s^{-1} are suitable as a natural liner.

- Permeability (*k*) of the wetland media.
- Plan area requirement for BOD removal is 5 m² per person.

All constructed wetlands require a polishing filter following the secondary treatment stage when discharging to ground (Figure 1). The polishing filter can reduce micro-organisms, phosphorus, and nitrate nitrogen in otherwise high quality wastewater effluents. The polishing filter produces a high quality effluent. The advice provided above allows effluent from a polishing filter to discharge to ground provided the subsoil has a suitable percolation rate[3].

Constructed wetlands that discharge to surface water may only do so in accordance with a Water Pollution Act discharge licence. Where a local authority grants a licence for discharge to surface waters from a constructed wetland system, conditions and restrictions in relation to the following may need to be met:
- System design and construction.
- Emission limit standards.
- Monitoring of inlet and outlet.

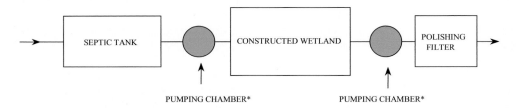

* Pumping is always required for intermittent filters. If the topography or the design permits, gravity
systems may be possible for constructed wetlands and polishing filters

Figure 1. Schematic of using a constructed wetland within a treatment system.

Installation, maintenance and monitoring

The design and installation of a constructed wetland are very important aspects that should be undertaken by a suitably qualified professional. The trial hole examination and subsoil classification carried out during site assessment will enable a decision to be made as to whether or not a natural or artificial liner is required. If an artificial liner is required then its installation should be certified.

The manner in which the treatment system is maintained after it is installed is of equal importance to ensure that the environment is protected on an on-going basis after the house is occupied. A general inspection of the system is recommended to ensure the effective operation of the wetland. The frequency and type of inspection should be determined taking into account the receiving environment and performance requirements. Initial wetland vegetation inspections should examine the viability of plant propagules. Subsequently,

---

[3] See Section 3 Of EPA 2000 for details of percolation tests and suitable rates.

wetland vegetation should be monitored, to determine percentage vegetation cover and height. These techniques can ascertain the development of vegetation and should be a routine part of operational wetland monitoring.

Current practice in Ireland

The guidance, which is produced by the EPA with regard to the use of CWs up to the present time, has focussed on the treatment of wastewater from single houses and small communities. Other applications of constructed wetland technology, for example the treatment of agricultural soiled water, have been identified in the Irish situation. However, no detailed guidance has been produced for such applications.

Wastewater from small communities and single house situations are successfully treated through the use of constructed wetlands. Among other developments in the area, proprietary modular reed bed systems have come onto the Irish market. While the performance of these systems has proven very favourable in other jurisdictions, data on the performance of these systems in Ireland is awaited.

Some local authorities are favourably disposed to the application of this technology; however, other local authorities are not convinced. Although guidance on the use of CWs has been provided (EPA, 2000) some local authorities view the technology as 'novel' and unproven in the Irish situation. To build on the successful use of CWs in the Irish situation the EPA will continue to foster research on the topic and will continue to provide guidance on the use of constructed wetland systems.

References

Cooper, P.F., G.D. Job, M.B. Green, and R.B.E. Shutes, 1996. Reed beds and constructed wetlands for wastewater treatment. Water Research Centre, Swindon, U.K, pp. 184.

Department of the Environment and Local Government, Environmental Protection Agency and Geological Survey of Ireland, 1999. Groundwater protection schemes. Department of Environment and Local Government, Environmental Protection Agency and Geological Survey of Ireland, Dublin.

Department of the Environment and Local Government, Environmental Protection Agency and Geological Survey of Ireland, 2000. Groundwater protection responses for on-site wastewater systems for single houses, Geological Survey of Ireland, Dublin.

EPA, 1999. Wastewater Treatment Manuals: Treatment systems for small communities, business, leisure centres and hotels. Environmental Protection Agency, Wexford.

EPA, 2000. Wastewater treatment manuals: Treatment systems for single houses. Environmental Protection Agency, Wexford.

Nutall, P.M., A.G. Boon, and M.R. Rowell, 1998. Review of the design and management of constructed wetlands. Construction Information and Research Association (CIRA), pp.268.

S.R.6, 1991. Septic tank systems: recommendations for domestic effluent treatment and disposal from a single house. Eolas, Dublin.

USEPA, 1992. Wastewater treatment/disposal for small communities manual. EPA/625/R-92/005. United States Environmental Protection Agency. Washington, DC.

USEPA, 2000. Onsite Wastewater Treatment Manual. EPA/625/R-00/008. United States Environmental Protection Agency. Washington, DC.

Cost effective management of soiled water from agricultural systems in Ireland

N. Culleton[1], E. Dunne[2,3,4], S. Regan[5], T. Ryan[6], R Harrington[7] and C. Ryder[8]
[1]Teagasc Research Centre, Johnstown Castle, Co. Wexford, Ireland
[2]Wetland Biogeochemistry Laboratory, Soil and Water Science Department, University of Florida/IFAS, 106 Newell Hall, PO Box 110510, Gainesville, FL 32611, USA
[3]Teagasc Research Centre, Johnstown Castle, Wexford, Ireland
[4]Department of Environmental Resource Management, Faculty of Agriculture, University College Dublin, Belfield, Dublin 4, Ireland
[5]Teagasc, Athenry, Co. Galway, Ireland
[6]Kildalton, Agriculture College, Piltown, Co. Kilkenny, Ireland
[7]National Parks and Wildlife, Department of Environment, Heritage and Local Government, The Quay, Co. Waterford, Ireland
[8]Engineering Services, Office of Public Works, Dublin, Ely Place, Dublin 2, Ireland

Abstract

Water quality is still a concern in areas dominated by agriculture. Pollution from agriculture can be from both point and diffuse sources. Ensuing water quality regulations will require changes in farming practices if water quality objectives are to be met. The objectives of this paper were to compare and contrast the monetary costs of soiled water management systems with alternative low cost systems in Ireland and outlined some advantages and disadvantages of each system. Our analyses indicate that on dry soil areas one week storage and auto-irrigation of soiled water was the cheapest method, in comparison to the other three storage and spreading options. The use of tractor and vacuum tank to spread soiled water, although it is the most widely used system for landspreading, is the most monetary expensive. Alternative low cost systems such as Integrated Constructed Wetlands (ICWs) were the second cheapest method for storing and treating soiled water on dry soil areas and in wet soil areas they were the cheapest. Results do not take into account the real costs of each system (environmental cost and benefit of each system); however our limited findings may provide some useful information for the comparison of each system.

Keywords: water quality, agriculture, soiled water, storage, landspreading, cost.

Introduction

Eutrophication is a water quality problem within agricultural watersheds around the world (Sharpley *et al.*, 2000; Stapleton *et al.*, 2000; Withers *et al.*, 2000). Increasing evidence suggests that intensive agricultural practices impact water quality (Rekolainen *et al.*, 1997). In the Republic of Ireland, nearly 15% of Irish lakes (between the years 1998 and 2000) were classified as eutrophic to hypereutrophic. Also, 30% of Irish river channels surveyed between those years were slightly to seriously polluted (McGarrigle *et al.*, 2002) of which agricultural practices were considered a significant contributor.

Pollution from agriculture can come from two possible sources. These include non-point and point sources. Non-point source pollution includes runoff from agricultural lands. This is of concern both in Europe and United States (Gburek *et al.*, 2000; Heathwaite *et al.*, 2000; Uusi-Kamppa *et al.*, 2000). In 1993, agriculture accounted for 50% of non-point inputs to surface waters in the UK (DEFRA, 2002). Recent increases in non-point source pollution in Northern Ireland are associated with agriculture operating at a P surplus of about 15 kg ha^{-1} yr^{-1} (Smith *et al.*, 2005).

Point source pollution is the issue being discussed in this paper and it is pollution that arises from a specific point within a landscape. It can include nutrient and contaminant losses from intensive livestock operations (Centner, 2000) such as mismanaged farmyard soiled water. Within Ireland and the UK, soiled water or farmyard dirty water from dairies includes farmyard washing, dairy parlour washings, effluents from silage pits, farmyard manure effluents, along with rainfall falling on open soiled farmyard areas (Brewer *et al.*, 1999; Cumby *et al.*, 1999). Pollutants in surface waters such as excessive concentrations of nitrogen and phosphorus, if lost from agricultural practices by point or non-point sources are regarded as lost resources (Braskerud, 2001); therefore it is in the interest of both farmer and the environment that every effort is made to prevent and control such losses.

Water quality regulations in Ireland

Two major European Union (EU) water quality directives: the Water Framework Directive (2000/60/EC) and the Nitrates Directive (91/676/EEC) are in the process of being implemented in Ireland and both have implications for Irish agriculture in terms of farming within water quality regulations. The Water Framework Directive (WFD) sets demanding quality standards for all waters and there is a precise time-scale for its implementation. The Nitrates Directive is designed to deal with specific problems such as water pollution caused by agriculture, and it must do so within the context of objectives and quality parameters set down in the WFD. The effectiveness of the legislation must be monitored and Member States are required to put more stringent measures in place, where water quality objectives are not met.

An Irish National Action Programme to implement the Nitrates Directive is implemented on a phased basis commencing on the 1st of January, 2005 and will operate for a period of four years. Implementation will be monitored on an ongoing basis by reference to water quality status and agricultural practices. The primary aims of the Action Programme are to reduce eutrophication of water resources caused and/or induced by nitrates and phosphates from agriculture and to prevent further pollution. The control and management of soiled water forms an important component of this National Action Programme.

Grassland based agriculture in Ireland

In Ireland, mainstream agriculture is grassland based and the objective of this paper is to assess the economic cost of managing soiled water, which is a by-product of grassland, based agricultural systems. Grassland based agriculture has two main features. The first is that approximately 90% of a dairy cow's diet is composed of grass, silage and hay, which is produced on the farm. The remaining ten percent of the diet consists of animal feed concentrates, which is imported to the farm. Thus, grassland agriculture contrasts with the intensive production of pigs and poultry, where the animals' diet is virtually all imported. The second feature is that all slurry (liquid animal excreta that has about six to eight percent dry matter content) and farmyard manure (decomposed dung and urine with straw and litter used as animal bedding) generated on the farm during livestock over wintering periods is recycled back on to grassland areas from which, hay and silage are harvested for animal feed. Table 1 illustrates the importance of slurry in standard Irish fertiliser recommendations for first cut silage. Significant savings of up to € 80 per hectare are made when slurry generated on the farm is recycled onto grassland areas.

If slurry is not stored properly during animal housing periods, DM content can be as low as two to three percent, which is somewhat similar to the DM content of soiled water. The Rural Environmental Protection Scheme (REPS) and the EU Nitrates Directive contain strict rules governing both the storage requirements and spreading of slurry/farmyard manure to grassland areas. These include:

- Periods when land application of manures are prohibited, while the closed period for spreading of slurry varies for different parts of Ireland. It generally extends from the 1st of October until January 15th. Spreading of farmyard manure is prohibited between the 1st of November and the 31st of January.
- Livestock slurry storage capacity must be sufficient for the full livestock housing period with specified minimum regional storage capacities ranging from 16 to 20 weeks.
- Soiled water may be landspread throughout the year provided weather and soil conditions are suitable. A minimum storage capacity of seven days is required for systems that landspread with an automatic irrigation system.
- Manures may not be applied to land where ground slopes are steep and where, taking into account factors such as proximity to surface water bodies, soil condition, ground cover and rainfall, there is significant risk to water quality.
- Manure application to land is prohibited under the following conditions:
  - If land is likely to flood or is waterlogged
  - When heavy rain is forecasted within 48 hours
  - Where land is frozen or snow-covered.
- Livestock manure including soiled water may not be applied to land within:
  - Five metres of a surface water body
  - Fifteen metres of exposed cavernous (karstified) limestone or karst limestone features such as swallow holes and collapsed features

Table 1. Nitrogen, phosphorus and potassium recommendations for first cut silage in late May/early June, with and without slurry application.

Slurry application t ha⁻¹	Total nitrogen kg ha⁻¹	Total phosphorus kg ha⁻¹	Potassium kg ha⁻¹
33	95	10	8
None	125	30	150

- Fifty metres of a borehole, spring or well used for the abstraction of drinking water for human consumption
- Specified areas around designated groundwater source protection zones.
- Slurry must be applied to land using tractor and vacuum tank that has an inverted splashplate, band spreader or trailing shoe, soil injection or soil incorporation methods associated with it.
- Livestock manure, farmyard manure, and other organic fertilisers may not be applied to land in quantities exceeding 50 tonnes per hectare at any one time.

Soiled water

An area that has not received sufficient attention to date is the storage and management of soiled water (which is often called "farmyard dirty water") on farms. In 1998, a desk study was conducted and it estimated that nearly 20 million tonnes of soiled water is produced on dairy farms annually (Brogan *et al.*, 2001). Soiled water produced on dairy farms can vary in nutrient content and volumes produced (Cooper *et al.*, 1996) both within and between farming practices (Hubble and Phillips, 1999). The environmentally safe management of soiled water remains one of the major challenges facing grassland-based agriculture in Ireland and there is a scarcity of information on soiled water produced on Irish dairy farms. Table 2 summarises some volume data of soiled water produced on farms in several countries. The amounts of soiled water produced are somewhat variable. Such variability suggests that management methods of soiled water may also be variable.

Brewer *et al.* (1999) showed that the volumes of dirty water produced on dairy farms in the UK were not related to a simple head count of dairy cows. Factors such as rainfall and size of soiled outdoor yard area had a major influence on volumes of soiled water produced. Soiled water along with varying in volume can also vary in polluting strength. Table 3 summarises some of the characteristics of soiled water from various dairy farms in Ireland, UK and USA. The variability in nutrient content and volumes of soiled waters produced indicate that on many dairy farms there maybe considerable scope for reducing loads by more effective management of waters.

Methods for storage and treatment of soiled water

Storage and treatment of soiled water varies with farming practice. Thus, what is appropriate management on one farm may not be suitable or effective for another; therefore a site-specific approach is both appropriate and required. Our analyses of management methods for soiled

Table 2. Sources and volumes of soiled waters produced in animal based farming systems.

Soiled water type	Mean $m^3 d^{-1}$	Min. $m^3 d^{-1}$	Max. $m^3 d^{-1}$	Location	Study
Dairy	7.2	1.2	19	Ireland	Dunne, 2004
Dairy	6.3	0.5	19	UK	Brewer et al., 1999
Livestock wastewater	10	-	103	USA	Knight et al., 2000

Table 3. Concentrations of water quality parameters in soiled water from dairy farms.

Soiled water type	TSS mg l^{-1}	BOD$_5$ mg l^{-1}	NH$_4^+$-N mg l^{-1}	PO$_4^{3+}$-P mg l^{-1}	K mg l^{-1}	Location	Source
Soiled water from dairy	910	2787	46	19	-	Ireland	Dunne, 2004
Yard runoff from dairy unit	745	2713	88	34	801	Ireland	Ryan, 1991
Irrigated soiled water from dairy	1838	2077	92	17	210	Ireland	Ryan, 1991
Soiled water from dairies	-	6593	457	415	1175	UK	Cumby *et al.*, 1999
Soiled water from dairy unit	1645	2683	7.7	-	-	USA	Schaafsma *et al.*, 1999
Soiled water from dairy unit	1284	2683	7.7	-	-	USA	Newman *et al.*, 2000

water are divided into several scenarios: those management practices used on farms located on wet soils, with poor permeability and those used on farms located on dry soils that have free draining soils. The functional differences in soil type influence the length of time required for animal housing during wintering periods and hence, impact soiled water management. Table 4 outlines some of the management options that maybe used. Dairy farms located on dry soils do not require long periods for storage, in comparison to farms located on wet soils, as they present more frequent opportunities for land spreading of soiled waters. In farms located on wet soils long-term storage is required because often times land is too wet for travelling with a slurry tanker to spread soiled water. Also, if spreading is carried out on wet soils before heavy rainfall events there is a likelihood of nutrient loss by surface runoff.

Option 1 in table 4 is the use of integrated constructed wetlands (ICWs) to store and treat soiled water. These biological systems are a relatively new method of managing such waters in Ireland. Table 5 considers some of the potential advantages and disadvantages of an ICW to store and treat soiled water. Another method for soil water management includes auto-irrigation systems (option 2; dry soils; Table 4), which are widely used in dairy farms across Ireland. The advantages of such systems to spread soiled water to grassland areas include: convenience and ease of operation, soil compaction due to heavy machinery is avoided, and

Table 4. Management options for soiled water on dry and wet soils in Ireland.

Options	Dry soils	Options	Wet soils
1	Integrated constructed wetland (ICW) for storage and treatment	1	ICW for storage and treatment
2	One week storage and auto-irrigation to grassland	2	Clay lined earthen bank tank (16 weeks storage) and umbilical system for spreading to grassland
3	Two weeks storage and tanker spreading to grassland	3	Clay lined earthern bank tank (16 week storage) and tanker spreading to grassland
4	Four week storage with an artificially lined earthen bank tank and tanker spreading to grassland	4	Artificially lined earthen bank tank (16 weeks storage) and tanker spreading to grassland

running costs are low. However, there are some disadvantages, which have become more apparent in recent years. Very often grassland areas allocated to auto-irrigation spreading are too small and often times land becomes saturated, which can result in surface run-off of land spread waters that can cause diffuse pollution to nearby water courses. Soil surface ponding within crop can also lead to reduction of grass growth. Leaching of nitrates to groundwater is also possible on light sandy or karstic soils. Best management practices for auto irrigation systems include:
• Correct adjustment to give the recommended application rate.
• Monitoring irrigator movement to avoid multiple applications to the same spreadland area.
• Installation of a trigger device to stop pumping and provide a warning when the irrigator reaches the end of its run in order to prevent soil surface ponding.

Short-term storage and spreading of dirty water on land by tanker (option 3; Table 4) is also a system that is widely practiced on farms, which are located on relatively dry soils. Storage of soiled water can be in a reinforced concrete structure and/or an earthen bank tank. The storage period can last from one to four weeks. The advantages of such systems are:
• Initial costs are low
• Post storage, nutrients in soiled water can be quickly recycled back to grassland areas.
The disadvantages include: short storage periods that may necessitate the spreading of soiled water on saturated soils. This can ultimately lead to soil compaction, surface ponding, surface runoff, which have negative for agronomic performance and environmental quality. More storage is needed in these circumstances.

On wet or heavy soils, the opportunities for spreading soiled water during long periods is not present in comparison to dry soils, with spreading restricted to about 20 weeks of the year. Therefore, adequate storage is critical during these periods of non-spreading. Earthen bank tanks (options 2 and 3; wet soils; Table 4) are often used to store large volumes of soiled water and currently the Department of Agriculture and Food approves earthen banks tanks, providing they are synthetically lined. However, synthetic liners are expensive; therefore earthen bank tanks that are lined with compacted clay are sometimes constructed as alternative low cost storage systems. Research undertaken by Teagasc and University College Dublin suggest that if designed and constructed appropriately, these systems provide safe storage of both slurry and soiled water (Scully *et al.*, 2002). Specifications and protocols for the construction

Table 5. Advantages and disadvantages of using an ICW to manage soiled water.

Advantages	Disadvantages
• Low operating costs	• High capital costs relative to other present management practices
• Easy to manage	• Land intensive ($\approx$ 1 hectare)
• Environmentally robust	• Short term loss of nutrients to grassland areas
• Creation of wild life habitat	• Planning permission required
• Suitable for wet and dry soils	• Discharge license required
• Potential for water conservation	• Initial site assessment required
• Long term storage of nutrients	• Professional design and construction required

of clay lined earthen bank tanks are being written and it is hoped that these systems will shortly be approved by the Irish Department of Agriculture and Food such that they can provide farmers with a low cost storage facility for such waters.

Management of soiled water: how much does it cost?

In the options that were previously considered, the ICW option had the lowest capital cost and the automatic irrigation system had the lowest annual costs. Both capital and annual costs can vary widely. In order to take this into account a ten year cash flow analysis was carried out. The results are presented for various grant and tax rate scenarios in Ireland. In each case, the auto-irrigation system was cheapest, with the ICW being the second cheapest (Table 6). Although tanker application of soiled water to grassland areas is the most widely used land application system in Ireland, this option had one of the highest annual costs. These arise from the associated labour and tractor running costs. However, capital costs are reduced when both tanker and tractor have other on farm uses. All of the systems shown in table 6 attract a government grant except the ICW option. These grants provide for a standard rate of 40% for fixed structures such as concrete tanks and 20% on mobile equipment such as vacuum tankers. The maximum investment limit is €75,000. The marginal rate of taxation (25% or 47%) at which individual farmers are liable had a significant bearing on costs (Table 6). Farmers with a low or no taxable income can least afford to invest in conventional structures and equipment and are increasingly looking at 'low cost' alternative systems such as clay lined earthen bank tanks as potential management options. Capital investment can be written off against the tax liability for a seven year period. We have not costed in monetary terms any possible environmental costs or benefits associated with any of these systems. Thus, our assessment of the "real costs" is very limited.

When conditions for landspreading are poor such as wet soil areas, the storage requirement may increase to 16 weeks or more. This has a corresponding increase in capital costs (Table 7). The capital cost of the ICW option remains similar to that option in dry soil areas and is unaffected by poor soil conditions. As concrete tank storage costs become prohibitive at the higher storage capacities only 'low cost' storage options are included in Table 7. The clay lined earthen bank tank, when appropriate has the lowest capital cost for this level of storage. The use of an artificial liner such as a high density polyethelene liner increases the capital cost three fold. In general, the annual costs varied widely between systems (Table 7). The wetland option has a much lower annual cost than any other storage/spreading system, mainly due to the low annual licensing and monitoring costs. However, these costs maybe subject to change. Spreading by contractor is assumed in view of the requirement for specialised high output low ground pressure equipment given the likelihood of small timescales for landspreading under wet soil conditions. When capital and annual costs are accumulated over a 10-year basis at various levels of taxation and grant-aid the wetland option is the cheapest option even though it is not subject to grant aid (Table 7).

Conclusions

Agriculture continues to contribute to the eutrophication of water resources in Ireland. Nutrient and contaminant loss from agriculture can be from both point and diffuse sources with point

Table 6. Installation costs and incurred costs annually and on a 10-year cash flow basis for four storage and treatment options of soiled water on dry soils at a farm-scale. It includes capital, annual and 10-year cash flow costs per head of livestock at various levels of taxation and grant-aid. All costs are given in Euro (€).

		Option 1	Option 2	Option 3	Option 4
	Method of storage	Integrated constructed wetland	Concrete tank (one week storage)	Concrete tank (two weeks storage)	Artificially lined earthen bank tank (four weeks storage)
(€)	Method of treatment	Integrated constructed wetland	Automatic irrigation system	Own tanker (50% use for soiled water)	Own tanker (50% use for soiled water)
Capital cost per head of livestock			105	148	108105
Annual expenses[1]			13	5.6	18.02 20.99
10 yr cash flow (per head)[2] No grant/low tax[4]			198	176	236242
10 yr cash flow (per head)[2] No grant/high tax[5]			136	125	167158
10 yr cash flow (per head)[2] Capital grant[3]/low tax[4]			198	123	209214
10 yr cash flow (per head)[2] Capital grant[3]/high tax[5]			136	87	148151
Grant for storage			No	Yes	Yes Yes

[1] Includes estimated costs of integrated constructed wetland (ICW) discharge license and monitoring

[2] Includes annual expenses, tax allowances and interest borrowed capital - borrowed over seven years at five percent.

[3] Capital investment grant rate is 40% (20% on mobile equipment such as tankers, subject to maximum investment level of €75,000)

[4] Low tax rate is 25%

[5] High tax rate is 47%.

source loss of soiled water from farmyard areas being a potentially acute source of pollution. As legislation is becoming increasingly stringent, there is a greater need for appropriate management of soiled water generated in farmyard areas to help prevent and control the impacts of agriculture on water quality. The appropriate management of soiled water presents a major challenge, as large storage capacities and robust systems for treatment are generally required. Minimisation of soiled water, through good farmyard design and management is the first step towards prevention. However, not all soiled water produced on farms can be cost effectively minimised; therefore pollution control measures are required and these include the several soiled water storage/treatment options outlined.

Our findings suggest that on dry soil areas, auto-irrigation of soiled water is the cheapest system to operate. However, there are environmental concerns associated with such an approach. The use of tractor and vacuum tank although the most widely used system for

Table 7. Installation costs and annually incurred costs on a 10-year cash flow basis for four storage and treatment options for soiled water on wet soils at a farm-scale. It includes capital, annual and 10-year cash flow costs per head at various levels of taxation and grant-aid. All costs are given in Euro (€).

		Option 1	Option 2	Option 3	Option 4
	Method of storage	Integrated constructed wetland	Clay lined earthen bank tank, 16 week storage	Clay lined earthen bank tank, 16 week storage	Artificially lined earthen bank tank, 16 week storage
(€)	Method of treatment	Integrated constructed wetland	Umbilical pipeline system (contractor)	Tanker spread by contractor	Tanker spread by contractor
Capital cost per head		105	62	62	190
Annual expenses[1]		13	22.24	28.88	31.44
10 yr cash flow (per head)[2] No grant/low tax[4]		198	223	273	408
10 yr cash flow (per head)[2] No grant/high tax[5]		136	158	193	288
10 yr cash flow (per head)[2] Capital grant[3]/low tax[4]		198	223	273	339
10 yr cash flow (per head)[2] Capital grant[3]/high tax[5]		136	158	193	240
Grant for storage		No	Yes (equipment only)	Yes (equipment only)	Yes

[1]Includes estimated costs of integrated constructed wetland (ICW) discharge license and monitoring
[2]Includes annual expenses, tax allowances and interest borrowed capital - borrowed over seven years at five percent.
[3]Capital investment grant rate is 40% (20% on mobile equipment such as tankers, subject to maximum investment level of €75,000)
[4]Low tax rate is 25%
[5]High tax rate is 47%.

landspreading soiled water was one of the most expensive. Integrated constructed wetlands were the second cheapest method for storing and treating soiled water on dry soil areas, whereas in wet soil areas, ICWs were the cheapest option. Thus, wetlands compared favourably under both wet and dry soil conditions even though they were not subject to grant aid.

References

Braskerud, B.C., 2001. Sedimentation in small constructed wetlands: retention of particles, phosphorus and nitrogen in streams from arable watersheds. Ph.D. dissertation. Agricultural University of Norway.

Brewer, A.J., T.R. Cumby and S.J. Dimmock, 1999. Dirty water from farms II: treatment and disposal options. Bioresource Technology **67** 155-160.

Brogan, J., M. Crow, G. Carty, 2001. Developing a National Balance for Agriculture in Ireland: A Discussion Document. Environmental Protection Agency, Ireland

CEC (Council of the European Communities), 2000. Council Directive of 23 October 2000 establishing a framework for community action in the field of water policy (2000/60/EC). O.J. L 327/1.

Centner, T.J., 2000. Animal feeding operations: encouraging sustainable nutrient usage rather than restaining and proscribing activities. Land Use and Policy **17** 223-240.

Cooper, P. F., G.D. Job, M.B. Green and R.B.E. Schutes, 1996. Reeds Beds and Constructed Wetlands for Wastewater Tratment. WRc, Swindon.

Cumby, T.R. A.J. Brewer, and S.J. Dimmock, 1999. Dirty water from farms I biochemical characteristics. Bioresource Technology **67** 155-160.

DEFTA, 2002. The Government's Strategic Review of Diffuse Water Pollution from Agriculture in England: Agriculture and Water: A diffuse pollution review. Department for Environment, Food and Rural Affairs, London.

Dunne, E.J., 2004. Wetland systems to mitigate contaminant and nutrient loss from agriculture. Ph.D. dissertation. University College Dublin, Ireland.

European Communities, 1991. Council Directive of 12 December 1991 concerning the protection off waters against pollution caused by nitrates from agricultural sources. (91/676/EEC). O.J. L 353.

Gburek, W. J., A. N. Sharpley, L. Heathwaite, 2000. Phosphorus management at the watershed scale: a modification of the phosphorus index. Journal of Environmental Quality **29** 130-144.

Heathwaite, L., A. Sharpley, and W. Gbruek, 2000. A conceptual approach for integrating phosphorus and nitrogen management at watershed scales. Journal of Environmental Quality **29** 158-166.

Hubble, I., and R. Phillips, 1999. Tasmanian dairy farm effluent management program. Journal of Cleaner Production **7** 167-168.

Knight, R L., V.W.E. Payne, R.E. Borer, J.R. Clarke J.H. Pries, 2000. Constructed Wetlands for livestock wastewater management. Ecological Engineering **15** 41-55.

McGarrigle, M.L., J.J. Bowman, K.J. Clabby, J. Lucey, P. Cunningham, M. MacCarthaigh, M. Keegan, B. Cantrell, M. Lehane, C. Cleneghan, and P. F. Toner, 2002. Water quality in Ireland 1998-2000. Environmental Protection Agency, Ireland.

Newman, J.M., J.C. Clausen, J.A. Neafsey, 2000. Seasonal performance of a wetland constructed to process dairy milkhouse wastewater in connectiant. Ecological Engineering **14** 181-198.

Ryan, M. 1990. Properties of different grades of soiled water and strategies of safe disposal. pp. 43-58. Proceedings of Environmental Impact of Landspreading of Wastes. Wexford, Ireland. 30-31 May, 1990, Teagasc, Dublin.

Schaafsma, J.A., A.H. Baldwin, and C.A. Streb. 2000. An evaluation of a constructed wetland to treat wastewater from a dairy farm in Maryland, USA. Ecological Engineering **14** 199-206.

Scully, H., P. J. Purcell, M. Long, T. Gleeson, E. O'Riordan, and S. Crosse, 2002. An evaluation of earth bank tanks for slurry storage. In: Proceedings of the Irish Agricultural Research Forum, Tullamore, 11-12 March 2002, p. 63.

Sharpley, A.N. and S. Rekolainen, 1997. Phosphorus in agriculture and its environmental implications. pp. 1-54. In: Phosphorus loss from soil to water, edited by Tunney, H., O. T. Carton, P. C. Brookes, and A.E. Johnston. CAB (Centre of agriculture and biosciences) International. Oxon.

Sharpley, A.N., B. Foy, and P. Withers, 2000. Practical and innovative measures for the control of agricultural phosphorus losses to water: an overview. Journal of Environmental Quality **29** 1-9.

Smith, R.V. C. Jordan, and J.A. Annett, 2005. A phosphorus budget for Northern Ireland: inputs to inland and coastal waters. Journal of Hydrology **304** 193-202.

Stapleton, L., M. Lehane, and P. Toner. (eds.)., 2000. Ireland's environment: a millennium report. Environmental Protection Agency (EPA). Wexford.

Uusi-Kämppaä, J., B. Braskerud, H. Jansson, N. Syversen, and R. Uusitalo, 2000. Buffer zones and constructed wetlands as filters for agricultural phosphorus. Journal of Environmental Quality **29** 151-158.

Withers, P.J.A., I. A. Davidson, and R. H. Foy, 2000. Prospects for controlling nonpoint phosphorus loss to water: A UK Perspective. Journal of Environmental Quality **29** 167-176.

Nutrient management in agricultural watersheds: A wetlands solution

Panel summary and concluding remarks

The symposium focused on the role of wetlands to improve water quality and to provide other ecological services within agriculturally dominated watersheds. Its primary objective was to disseminate information from international and national experiences to Irish researchers, regulatory agencies, engineers, non-profit organisations, farmers and the public on wetland functions, values and uses within agricultural watersheds.

On the final day a panel consisting of seven members was arranged to discuss some of the main components of the symposium. This chapter outlines some of the comments raised during this discussion, which addressed water quality, wetland nutrient cycling, functions and values of wetlands, constructed wetland management/operation, and policy considerations. Some issues for future consideration were also discussed. Panel members were:

- Donald Graetz, University of Florida, USA.
- Eddie Riordan, panel chairman, Teagasc, Grange, Ireland.
- Jan Vymazal, ENKI, Czech Republic.
- Noel Culleton, Teagasc, Johnstown Castle, Ireland.
- Padraic Larkin, Environmental Protection Agency, Ireland.
- Sean Regan, Teagasc, Athenry, Ireland.

Water quality

The general overview was that the majority of Irish freshwaters were not seriously polluted with about 70% of Irish waters being of satisfactory water quality status. However, 30% of Irish waters are degraded to some extent. Of that 30% the biggest water pollution problem is eutrophication; with its origins being attributed to agricultural, municipal and industrial sources. Agriculture is often cited as a major contributor.

Legislative changes in terms of agricultural practices and water quality are becoming increasingly stringent. The regulatory environment puts increased pressure on farmers who must change their farm practices to achieve water quality standards at the field-, farm- and watershed-scale. This is set against a background of change in European agriculture, which will see a decline in the number of full-time farmers and an increase in intensity and scale of their operations to remain competitive in the more open world marketplace. The increased operational scale on animal production farms will require increased facilities for the storage and management of animal manures and dirty waters generated in farmyard areas. Thus, there is interest in alternative low-cost farmyard systems as potential best management practices (BMPs). These can include both the building of new structures with the objective of pollution prevention, and non-structural BMPs, which can help control pollution. Three approaches to preventing and controlling pollution from farmyards have generated significant interest in Ireland. These are (1) earthen bank lagoons for slurry /effluent storage, (2) out wintering pads as an alternative housing system for bovines, and (3) wetlands for the treatment of farmyard dirty water.

Point source pollution

Farmyards, in particular, were identified as one of the major sources of point pollution within agricultural watersheds. To date, however no systematic evaluation of their pollution potential has been undertaken. The dirty water generated in farmyard areas has and continues to be, an area of concern because if unmanaged (or managed inappropriately) these waters have a high polluting potential since more than 100,000 farmyards exist in Ireland.

A functional definition of what farmyard dirty water is and is not is important, such that appropriate measures can be undertaken for the treatment and management of those waters. During the symposium such water was called "farmyard dirty water," "soiled water," "yard water," "yard runoff" and "agricultural wastewaters." All terms incorporated similar components, which included dairy parlour washwaters, rainfall on open concrete yard areas (and in some cases rainfall on roofed areas), open concrete yard washings, silage effluent, and manure effluent. From the management perspective, the characterisation of such waters should include a range of water quality parameter concentrations that have both minimum and maximum levels. A range of water quality parameter concentrations is more appropriate than absolute values as water quality parameters are variable depending on farmyard. It was agreed a broad range of values (quality and quantity) would facilitate site-specific variability, while being sufficiently quantitative such that farmyard dirty water could be characterised for management solutions. The legislative characterisation of these waters was not discussed.

In terms of BMPs, waste minimisation/prevention is a first principle. However, it is inevitable that not all waters, which have a polluting potential can be minimised in a cost effective manner due to spatial and temporal variability at the farm-scale. Therefore, both structural and non-structural BMPs are required, some of which may potentially include wetlands. With an integrated approach of both structural and non-structural BMPs nutrient and contaminant loss from agriculture can be prevented and/or controlled.

Non-point source pollution

Evidence for nutrient and contaminant loss from diffuse sources such as agricultural fields and agricultural watersheds was presented during the symposium. Evidence was presented using case studies from Ireland, Norway, Czech Republic, Spain and the USA. Forms of diffuse losses included nutrients such as nitrogen (N) and phosphorus (P) in addition to the loss of clay and suspended solids. Evidence illustrating the use of wetlands to retain nutrient and contaminant loss from diffuse sources was presented. Diffuse losses from agriculture are often difficult to eliminate. Therefore, alternate measures need consideration for pollution prevention and control. The use of constructed wetlands could be one approach.

The role of wetlands within the Irish agricultural landscape

There was general agreement that wetlands do have an important role to play within the Irish agricultural landscape, both in terms of improving water quality, and providing other ecological services such as hydrological buffers and ecological habitats. However, it was outlined that to date, regulatory agencies in Ireland are cautious about using wetlands, due to the perceived absence of scientific data. This has resulted in somewhat mixed and negative attitude toward the use of wetlands for the control of nutrient loss from farmyards in Ireland.

It was noted that there have been over ten international conferences on the use of wetland systems for water pollution control. Therefore there is a large international information base on these systems. It was also emphasised that ecosystems such as wetlands and grasslands have limitations. Given the dynamic nature of biological systems that include wetlands, grasslands, tillage crops and natural areas, simulating ecosystem structure and function and optimising those functions for some purpose i.e. water quality improvement, is challenging. However, it was agreed that what can be done is to conduct research to understand system characteristics and to develop models to set ecosystem limits of what potentially can and cannot be achieved using natural systems to improve water quality. An important component of this effort would be the development of long-term data sets of wetland performance.

Anecdotal evidence regarding the misuse of wetland systems for the treatment of polluted water was raised. However, the need to reflect on the efficacy of wetlands in totality rather than illustrate the mismanagement of these systems, was emphasised.

Water quality regulations such as the Water Framework Directive (WFD) require a more holistic and integrated approach to the management of water resources on a watershed-scale in comparison to previous EU legislation. Thus, the use of wetlands within watersheds is appropriate since by their very nature wetlands are integrated ecosystems exhibiting terrestrial and aquatic traits. The beneficial use of wetlands, in terms of improved water quality within an agricultural watershed was shown in the Anne Valley, Co. Waterford both during paper presentations, and in the post-symposium field trip. In addition, the use of wetlands (constructed and natural, which were either passive or actively managed) was shown to have positive effects for restoring hydrology and retaining nutrients and contaminants within agricultural watersheds in Europe and the USA. There were several approaches, with no one application similar to the next, illustrating the importance and need for a site-specific approach.

It was generally agreed that wetlands have a positive role within the Irish landscape. However, there was little discussion and or clarification on the use of performance measures, as a tool for successful management of wetlands to help improve water quality. It was mentioned that the use of performance criteria would help establish priorities for wetland use. This could provide a template for decision making such that there would be a feedback mechanism for programme improvement in order to apply lessons learned from experiences gained.

Differences in perspectives

It was evident from the symposium that water quality problems were similar irrespective of global geographic location. Changes in land use around the world have, are and will continue to impact water quality. It was also emphasised that wetland structure and function regardless of geographical location are similar because wetland ecosystems all consist of soil, water, vegetation and biological organisms. Although water quality problems are similar, unique issues dependent on climate, geology, hydrology, soil and land use which may depend on geographical location often arise. For example, P biogeochemistry in Florida is unique as soils are generally sandy and have a low P binding capacity. Water tables are typically shallow and are interconnected with freshwater systems. Land uses include intensive livestock, crop production and a rapid urbanising landscape. Climate is somewhat extreme with high rainfall (about 1000 mm yr^{-1}), most of which falls in summer months during intense rainfall events. Rainfall during a 25 year 24 hr storm event may be up to 20 cm. The ultimate P sink of landscapes in Florida are wetlands and lakes, as there is very little P retention within the terrestrial landscapes. In comparison, there is greater P retention capacity within Irish terrestrial uplands, wetlands and streams. This again is dependent on geology, hydrology, soil and land use. In Ireland, soils predominantly originate from glacial or limestone tills, where soils have high contents of silicate clays, iron oxides and are typically acidic in nature. Due to higher clay and iron oxide contents in Irish soils relative to Floridian soils, Irish soils typically have a higher capacity to store and retain P, as it was shown that P retention in mineral soils is governed by aluminium and iron content. In Ireland, about 50% of the landscape is drained by nine rivers and there are over 6,000 lakes. The total land area is underlain by bedrock where groundwater is stored, in addition to groundwater storage in sand and gravel aquifers. Over 60% of the land area is agricultural, of which 80% (3.5 million hectares) is grassland used for the production of milk and beef, whereas in Florida, agriculture uses about 30% of the total land area with 33% (1.4 million hectares) of that allocated to pasture. Rainfall in Ireland is similar in annual totals (750-1,200 mm) to Florida. However, the patterns of that rainfall are dramatically different. Average hourly rainfall amounts in Ireland are quite low, ranging from 1 to 2 mm.

In Florida, wetlands are often used on a large landscape-scale such as the use of storm water treatment areas (STAs) in South Florida. Evidence for their successful operation with a sole objective to reduce P from incoming agricultural drainage waters was discussed. The scale of wetland application to date in Ireland and some other European countries is small in comparison. Case studies were used to illustrate application at the farm-, field- and small watershed-scales.

Integrated constructed wetlands

The ICW session, which was specific to Ireland, highlighted the main thrust for using wetlands to help improve water quality and provide ecological services. However, other presentations outlined that there were other wetland approaches being used in Ireland. The initial concept of the ICW approach with aspects of site assessment, planning and performance was illustrated. It was indicated that this wetland initiative was first undertaken in the 1990s, where wetlands were constructed to treat farmyard dirty water and provide ecological services in Co. Waterford. It was constantly emphasised during the symposium that there is a need for a nationally agreed to protocol for this approach. Information was presented outlining a draft template for that protocol. To date, there is no formal publication of the protocol. It is anticipated that an Irish protocol will be published in 2005.

Biogeochemistry

During the N and P biogeochemistry sessions an understanding of the complex transformations and translocation of nutrients within wetland systems was gained. The environmental conditions, which regulate wetland N and P dynamics were also highlighted.

Some understanding of wetland biogeochemical processes is important for successful wetland operation. Biogeochemical processes interact between and within wetland components that include wetland soils, overlying waters, soil porewater, accumulating sediments, vegetation, and microbial communities. The knowledge-base presented at the symposium suggests that there is ample understanding of wetland biogeochemistry, such that these systems could and are being tweaked, in order to provide functional roles within landscapes.

It was highlighted that wetland systems have characteristics of both terrestrial and aquatic systems, but also characteristics that are inherent to wetlands. Several differences outlined included: the proximity of aerobic/anaerobic soil layers, hydrological and nutrient cycling functions of wetland vegetation, importance of microbially-mediated nutrient transformations, and abiotic nutrient translocations.

Wetland case studies presented evidence of N and P storage. Biogeochemical processes responsible included nitrification/denitrification, sorption and precipitation, in addition to the accumulation and accretion of organic matter. Long term retention of N was primarily governed by nitrification/denitrification cycles, with retention of P being primarily controlled by the accumulation of organic matter. Accretion rates of the accumulating organic mater (which typically has different physicochemical characteristics from the underlying mineral soil) can range in millimetres, and in some heavily impacted wetlands, accretion rates maybe in centimetres. Comments outlining past experiences from the USA suggested that the retention of nutrients, such as P, by long-term processes, can be greater than 90% during a 30 to 40 year operational period.

There were some questions raised concerning the fate of accumulated organic material within a constructed wetland, which may have high concentrations of organic matter and nutrients either after several years of wetland operation or post wetland operation. Depending on wetland discharge water quality, accumulated material can be excavated out of the constructed wetland system and applied to grassland areas as an organic fertiliser. This approach is similar to conventional land spreading of animal manures. As with manure, characterisation of this material is necessary prior land spreading, so that the nutrients applied, reflect crop requirements. It is important to note that the physicochemical characteristics of this material is very different to that of the original incoming farmyard dirty water, as farmyard dirty water is transformed by physical, chemical and biological processes within the wetland system. The accumulated material may also be applied or used for market garden crops, and as a soil conditioner, where soils are depleted in organic matter content.

Finally, as P is often the most land limiting of nutrients, a wetland solely designed and constructed for P storage and retention will have a larger land area requirement, in comparison to a wetland that is designed and constructed to retain nitrate, organic material and/or suspended material.

Policy and management

Policy and regulatory management of constructed wetland systems for the purpose of water quality improvement were discussed. An important comment noted several times was the lack of consultation between the Irish regulatory agencies and those involved in the planning, construction and operation of constructed wetlands. It was suggested that this may have been an important factor in the divisive national debate on wetlands. It was agreed that this development and the real or apparent differences between scientists, policy makers and end users has not been helpful.

The symposium showed that there is a considerable depth of knowledge available on wetland functions and uses, both internationally and in Ireland. Therefore, the lesson for Ireland is that research, operation, and monitoring the efficacy of each system should be rigorously assessed and the general components for the success of each system identified using a strategic management approach. It is the responsibility of scientists, engineers and those who operate such systems to provide the relevant information and for that information to be validated. This would help to address information gaps on a site-specific basis. However, these groups alone cannot provide all the answers. Therefore, based on an assessment of the information provided, policy makers and regulators must make decisions on such systems, accepting that there is always uncertainty with any management system. There is a need for policy and regulatory decisions that allow or provide a template for the accumulation of new information for future decision making i.e. a regulatory approach of "presumed compliance." For this to happen, a nationally agreed protocol for the design, construction and operation of a constructed wetland is required. In theory, a farmer that had a constructed wetland designed, built and subsequently operated to a nationally agreed protocol (which was flexible enough to incorporate site-specific conditions) would be presumed compliant with legislative regulations. Regular assessment (annual, bi-annual or quarterly) of performance criteria is a necessary component of this approach. The presumed compliance approach is used elsewhere around the world to help initiate new agricultural management systems and accumulate information, which can help future decision making.

A systematic approach for continual improvement of wetland design and operation by learning from programme outcomes is also beneficial and appropriate. This approach employs the concept of "adaptive management". Adaptive management has been used in natural resource management since the 1970s. It is most effective when alternative practices and policies are experimentally compared and subsequently evaluated. Such an approach includes several components: education/training programmes, and the use of demonstration projects. The information gained though monitoring and assessment of such projects is used in future projects, so that decisions are made based on the lessons learned from previous experiences. In general, such an approach should rely on a good knowledge base of "sound science," coupled with long term monitoring of system efficacy across environmental conditions. An integral component of this approach also includes system modeling, and the use of those long term datasets to help model system responses.

Future considerations for Ireland

In terms of water quality in agricultural watersheds and the use of wetlands to help improve water quality within agricultural watersheds, aspects of future research and initiatives that were identified included:

- Effects of extreme rainfall events on nutrient and contaminant transport at the farm-, field-, and watershed-scales.
- Cumulative impacts of agricultural practices on water quality within watersheds during extreme and steady-state events.
- Constructed wetland nutrient dynamics at farm-, field-, and watershed-scales within agricultural watersheds.
- Efficacy of constructed wetland systems during extreme rainfall events.
- Physical disturbances within watersheds that impact water quality, i.e. stream bank erosion, river channel instability, river sediment transport, elevated surface runoff, and agricultural soil surface ponding.
- Use of constructed wetlands to help restore/enhance river and stream systems within agricultural watersheds.
- Use of wetlands to provide ecological services within agricultural watersheds.
- Decision frameworks to help identify and quantify pollution problems and their resulting symptoms.
- The continued use of structural and non-structural BMPs to help mitigate contaminant and nutrient loss from agriculture.
- Frameworks to actively assess the effectiveness of BMPs.

Throughout the symposium, the need for an increase in consultation and participation in wetland issues as they relate to agriculture was emphasised. This should include increased stakeholder involvement and communication between regulatory and planning agencies, researchers, engineers, contractors, and most importantly the farmer/rural community. It was suggested that public-private partnerships may be helpful to increase effective communication between stakeholders. It was recognised that there is some collaboration between state agencies in an effort to produce a nationally agreed protocol for constructed wetlands.

It was agreed that alternative measures will be required to meet the 2015 water quality standards required under the WFD. There is a need to collate and synthesise the current body of knowledge on the use of constructed wetlands within agriculture and to use this information to validate and further improve future approaches and techniques.

Wetland ecosystems both natural and constructed are complex systems. To understand how wetlands function requires some knowledge of biology, hydrology, ecology, chemistry, engineering, and soil science. Thus, to construct a successful constructed wetland system that improves water quality and provides ecological services, interdisciplinary expertise is required. Interdisciplinary education and training programmes must be developed to provide information in a clear and easily understandable format for end users.

To conclude, future initiatives should incorporate a more holistic approach to using wetlands within agricultural landscapes. This approach will enable wetland functions and values be recognised by all stakeholders.

Authors index

Keyword index

Printed in the United States
by Baker & Taylor Publisher Services